CELESTIAL ENCOUNTERS ―――――――

Florin Diacu and Philip Holmes

CELESTIAL ENCOUNTERS

The Origins of Chaos and Stability

PRINCETON UNIVERSITY PRESS · PRINCETON, NEW JERSEY

All Rights Reserved

Library of Congress Cataloging-in-Publication Data

Diacu, Florin, 1959–
Celestial encounters : the origins of chaos
and stability / Florin Diacu and Philip Holmes.
p.   cm.
Includes bibliographical references and index.
ISBN 0-691-02743-9 (cl : alk. paper)
1. Many-body problem.   2. Chaotic behavior in systems.
3. Celestial mechanics.   I. Holmes, Philip, 1945–
II. Title.
QB362.M3D53   1996
521—dc20   96-108

This book has been composed in Times Roman

Princeton University Press books are printed
on acid-free paper and meet the guidelines
for permanence and durability of the Committee
on Production Guidelines for Book Longevity
of the Council on Library Resources

Printed in the United States of America by
Princeton Academic Press

10   9   8   7   6   5   4   3   2   1

To Marina and
Răzvan Diacu,
for their warmth
and love

In memory of Robert Montague Holmes,
1903–1995,
who, without knowing it, started his son on
this strange path

Tell me these things, Olympian Muses, tell
From the beginning, which came first to be?

Chaos was first of all

—Hesiod, Theogeny, II
114–116

It often happens that the understanding of the
mathematical nature of an equation is impossible
without a detailed understanding of its solutions.   —Freeman Dyson

Contents

---

* Indicates technical material

---

[*] Indicates technical material

Preface and Acknowledgments

THE GENESIS of this book is perhaps unusual. Florin Diacu (FD), the first author, is an émigré from Romania who now lives in western Canada, where he teaches mathematics at the University of Victoria. Prior to his departure for Germany shortly before the revolution of 1989, he earned his living as a school teacher. He worked in his spare time on the problems in celestial mechanics that would become his Ph.D. thesis, submitted to the University of Heidelberg. In 1992–93, one year after settling in Victoria, FD taught a graduate course on dynamical systems, during which he posted pictures of strange attractors and other chaotic objects on his office door. These prompted visits from students in various fields who wanted to talk about chaos. They frequently asked him to recommend a popular book. Because there seemed no obvious choice, FD thought of writing a book himself—an activity that would also help improve his English. In writing mathematical papers, one exercises only a limited vocabulary.

He had in mind a book for nonspecialists, aiming to describe the ideas and people that had most impressed him in his chosen field of research. He had taught young people with little mathematical knowledge, and so he might be able to convey, to a general audience, some of his ideas and the excitement he felt for the subject. Books such as James Gleick's *Chaos—Making a New Science* and Ian Stewart's *Does God Play Dice? The Mathematics of Chaos* had brought some of the current personalities and issues to the general public, but he felt that they were in some sense buildings without foundations. In spite of these attractive accounts, there still seemed to be room for a book that would go deeper into the history and mathematical foundations of a more limited but central problem in the field of chaos and dynamical systems theory: the *n*-body problem. Much of this history did not seem to be widely known. Even in the fine display devoted to chaos in the *Palais de la Découverte* in Paris, not far from the Sorbonne where Henri Poincaré taught, one finds the claim that the phenomenon was discovered in the 1960s.

While visiting Cornell shortly after his arrival in Montreal for postdoctoral studies, FD met Philip Holmes (PH). The latter was no expert in celestial mechanics, but he had been working on dynamical systems for twenty years. (Though one might say that celestial mechanics is a small part of dynamical systems theory today, this book will show that the roots of the latter subject lie in celestial mechanics.) After completing a first draft in early 1994, FD asked PH if he would read and criticize the rough text. The latter agreed, and subsequently returned a manuscript liberally annotated in red ink. At this point FD invited him to become the second author. PH was at first reluctant. What could he contribute, other than suggestions on style? But he too had not been entirely satisfied with published popularizations of recent discoveries in dynamical systems, and as time went on, he realized that he might have a contribution to make in explaining some of the mathematical ideas that form the heart of his own research. And so this book came to be.

The story of Henri Poincaré's discovery of chaos and the checkered history of its publication after it had won the prize offered by King Oscar II of Sweden and Norway has already been told in Ivars Peterson's fine book, *Newton's Clock: Chaos in the Solar System*. In retelling it in chapter 1 we make use of recent historical research and also attempt to give a deeper feeling for the theory of differential equations and the geometrical approach to their study that Poincaré invented. Peterson is primarily concerned with the history of developments in that part of astronomy relevant to the dynamics of the solar system. We shall be more interested in the mathematical foundations of *celestial mechanics* and of *dynamical systems theory* in general, which, as we shall see, are rather different creatures. In this chapter we introduce *phase space*, the mathematical universe that dynamical systems inhabit, and explain some of the tools for its analysis. We continue to develop them through chapter 2, as we describe the analytical, geometrical, and symbolic methods due to the American mathematicians George Birkhoff and Stephen Smale.

In chapter 3 we return to those problems of celestial mechanics that had been Poincaré's inspiration, describing the theory of singularities and collisions. Although these may seem unlikely events in the solar system in our lifetimes, they are important in the longer-term history of the universe and are crucial to a proper understanding of the structure of phase space for the n-body problem. (In fact, as we worked on this book in the summer of 1994, the fragments of comet Shoemaker-Levy struck Jupiter, and collisions were much in the news.) This chapter starts with Painlevé's lectures in Stockholm—he also had a Swedish connection—and carries us through the work of von Zeipel up to the most recent developments.

Chapter 4 is concerned with stability. We go back in time again to d'Alembert, Laplace, Lagrange, and Poisson. We then leave France to travel east to Romania and Russia to meet Spiru Haretu and Aleksandr Mikhailovich Liapunov. In the course of this journey we trace the successively more detailed approximate series solutions developed over two centuries and the contradictory conclusions regarding stability to which they led. The stability debate was not really settled until the 1950s and 1960s, and chapter 5 is devoted to the work that achieved this resolution. We describe the Kolmogorov-Arnold-Moser, or KAM, theory, one of the most influential (and certainly most quoted) bodies of results in classical mechanics. In doing so we see how chaos and stability are intimately linked in our mathematical models of the universe.

We have a twofold aim in this book. We wish first to relate some historical developments in celestial mechanics and dynamical systems theory, and in doing so, attempt to re-create the social and intellectual milieus in which the people responsible for them lived. Second, we hope to explain in some depth the mathematical ideas and methods that these pioneers left for us, and on which our own—far smaller—contributions have been built. This book is intended for anyone who has heard about chaos and wants to learn more about the theory and origins of the subject and the people who created it. In spite of chaos's more popular manifestations, including games and toys in airport gift shops, it is primarily a *mathematical* theory. We hope that a wide range of readers will find the stories that lie behind the ideas as interesting and exciting as we do, and that this will encourage them to probe into the more arcane aspects of the mathematics. Consequently, we have written neither a conventional historical study nor a survey of new ideas meant solely for the benefit of our fellow researchers and scholars. In fact, our approach may dismay some of them, and a brief explanation is in order.

What would our lives be without dreams and what would dreams be without imagination? In many respects, scientists are just as much dreamers as artists, composers, novelists, or poets. Like many of the latter, scientists launch their flights of fancy from the realities of the world. Mathematicians are no exception to this rule, but they differ from all other scientists in that their dreams are, at least in principle, 100% verifiable. Other scientists must rely on unsure hypotheses, unclear assumptions, or imperfect observations and experiments. After agreeing on basic axioms, the mathematician can build with precise notions and definitions and therefore whatever his imagination and skill constructs, stands on firm foundations. Every mathematician's vision is to build perfect edifices in concert with the spirit and requirements of his field. This is at once the great strength and a great limitation of mathematics: it is literally a world unto itself.

But mathematicians are still human beings, and the stimuli for their research and many of their ideas come from the world outside—the problems and phenomena of everyday life. No matter how dry and technical a theorem may seem, the impetus to state it and the ideas involved in its proof arose in some part from its creator's common experience, from seemingly casual conversations with friends and acquintances as much, perhaps, as from study of any particular "real" phenomenon. We hope to convey something of this interplay among chance events and meetings in mathematicians' personal lives, their insights, and the contributions they made. Mathematics is an international language, and its practice is a collective as much as a personal affair. Our story will take us to many different countries and cultures, and we will see how political and social factors have also helped mold the subject.

Any student struggling with a difficult proof or piece of analysis must have wondered where the great masters of the past got their ideas. Alas, those sorts of details rarely make their way into the published papers, or even into the historical literature. Sometimes they can be found in letters, diaries, and personal papers, but often they are lost when the person dies. We wanted not only to show how different discoveries were related "after the fact," within the larger world of mathematics, but also to identify their sources in the human world. This was fairly easy for those of our contemporaries who appear in the book. When the genesis of an idea or the circumstances of a proof were unclear, an E-mail message or a phone call would usually resolve the question (or perhaps elicit several contradictory resolutions). It was harder to uncover the stories of those who are no longer with us. The patina of time has obscured their features, and sometimes we could find out very little beyond the bare facts of dates and places.

In attempting to bring these figures to life, we have occasionally chosen to illuminate corners that lie beyond the light of historical research. While most of our story is firmly based on documented historical fact, we have created a few scenes and conversations for which there are no direct data. We have generally based these on analogies with similar, recorded events in the lives of the people concerned. Unlike the dreams of mathematicians, these of ours are unverifiable (but also unfalsifiable). We hope that they may help the reader to better appreciate mathematical creation, and ultimately the mathematics itself. In the notes that follow the main text, we clearly identify such passages, as well as provide sources for the better-documented events.

In choosing to write about the roots of dynamical systems theory from the standpoint of celestial mechanics and the n-body problem, we have intentionally limited ourselves to the mathematical foundations of the subject. We apologize in advance to readers who will not find their own specialties or applications here. Dynamical systems is such a rapidly growing field that it would, in any case, be impossible to provide a comprehensive survey within the confines of the present book. Alas, we have only finite time and energy!

Many people have helped us in this project: some contributed valuable information, others read the manuscript (or significant parts of it), curbing our excesses, making corrections, pointing out omissions, and forcing us to better explain ourselves. We are deeply indebted to Karl Andersson, Tom

Archibald, Vladimir Igorevich Arnold, Mark Bannar-Martin, Ilie Barza, Marc Chaperon, Alain Chenciner, Catharine Anastasia Conley, Scott Craig, Marina Diacu, Gabor Domokos, Robert Garber, Joseph Gerver, Robert Ghrist, Arlene Greenberg, Holly Hodder, Ruth Holmes, Martha Mary Jackson, Chris Jones, John Mather, Richard McGehee, Vasile Mioc, Jürgen Moser, Jaak Peetre, Chris Phillips, Moireen Phillips, John Phillips, Bill Reed, Donald Gene Saari, Jean-Claude Saut, Yasha Sinai, Steven Smale, Dana Schlomiuk, Norbert Schlomiuk, Carl Simon, Garry Tee, Stathis Tompaidis, Ferdinand Verhulst, Bogdan Verjinschi, Zhihong Xia, and Smilka Zdravkovska. We apologize to anyone we have inadvertently neglected to mention.

We would also like to thank our editor, Trevor Lipscombe, and his colleagues at Princeton University Press, especially Alice Calaprice, who painstakingly edited the manuscript; and we thank Chris Brest, who turned our rough sketches into clear figures.

Philip Holmes benefited from the support of the John Simon Guggenheim Memorial Foundation during part of the period in which this book was written. Finally, without E-mail and TeX, this long-distance project would have remained merely a dream.

Victoria, B.C., and Princeton, N.J.
April 1996

A Note to the Reader

P<small>ARTS OF THIS BOOK</small> are fairly technical, as we develop the fundamental mathematical tools necessary to give a clear and rigorous description of chaos. We have noted these sections in the text by adding an asterisk to the subheads. Those seeking a historical overview can skip such parts, at least on a first reading, and pick up the story later. These sections do not require more background knowledge, but they do demand more time and tenacity from readers unfamiliar with the ideas.

CELESTIAL ENCOUNTERS ──────────

1.

A Great Discovery—And a Mistake

It is probable that for half a century to come it will be the mine from which humbler investigators will excavate their materials.

—George Darwin, concerning Poincaré's masterpiece, *Les methodes nouvelles de la mécanique celeste*

Henri Poincaré pushed back his chair and stood up. He had to get out for a while. It was a beautiful spring afternoon: the sounds and smells drifting through his open study window made him restless and unaccountably optimistic. There was no other reason to feel encouraged, since his work seemed to be at a complete standstill. How could he believe what his calculations were telling him? Although each step of the argument followed logically from the previous one—after all, he had constructed them himself and checked them repeatedly—he could not grasp the whole. He had first tackled the problem ten years earlier and returned to it repeatedly, but a crucial insight was still missing. Until he found it, he would be unable to finish the final chapter of his book.

A break sometimes helps one overcome such obstacles. He would go for a walk. It would not matter where his steps took him. The important thing would be to free his mind and pay attention to the world around him. Paris at the end of the nineteenth century was a charming mixture of parks, history, cultural monuments, and romance, as it remains today. There would be plenty to see and think about as he walked. After an hour or two he could return to his desk, refreshed.

Although he possessed unusual gifts, Poincaré had for many years appeared to live like any other Parisian bourgeois. His acquaintances and colleagues saw him as a respected professor and member of the community, a loving husband, and an attentive father. Only within himself, when not consumed by research, was he aware of how difficult it was to divide his time among teaching, administrative duties, and his family.

A Walk in Paris

Jules Henri Poincaré was born on April 29, 1854, in Nancy, where his father was a respected physician. Nancy is the former residence of the dukes of Lorraine and the present government seat of the Meurthe-et-Moselle region. It lies in northeastern France on the Meurthe River and the Marne-Rhine Canal, 175 miles east of Paris. Both of Henri's parents were born in Lorraine, and the Poincaré family had lived in that area for some time. Jean Joseph Poincaré, an ancestor, was *conseiller au bailliage* (a judicial officer) in Neufchâteau, where he died in 1750. One of his sons, Joseph Gaspard Poincaré, taught mathematics at the Collège de Bourmont, nearby.

This distant mathematical connection did not prepare the family for Poincaré's arrival. He was a precocious child who rapidly surpassed his schoolmates in all subjects. The Franco-Prussian War interrupted his education, but during the occupation, from 1870 to 1873, he quickly became fluent in German without formal lessons. His fascination with mathematics began when he was fifteen. Once he had become absorbed in a problem, noise and activity in the room could not divert him.

In 1873 Poincaré graduated at the top of his class and entered the prestigious École Polytechnique in Paris, where he was able to follow the mathematics lectures without needing to take notes. His original concentration was mining and geophysics; in fact, he was admitted to the École National Supérieure des Mines in 1876, but he soon abandoned engineering and began doctoral studies in mathematics at the University of Paris in 1878, obtaining his Ph.D. in the amazingly short space of a single year. His thesis, "The Integration of Partial Differential Equations with Multiple Independent Variables," addressed a difficult technical question. It was to be followed by a flood of papers and books, which would define whole new areas of mathematics and change the course of others. It is amusing to note that he scored zero in the drawing examination for entry to the École Polytechnique and a special exception was necessary to admit him. The man who was to reintroduce a geometrical approach to dynamics was practically unable to draw a coherent picture.

It was the spring of 1897 when we joined Poincaré struggling with his problem. Approaching his mid-forties, a professor at the University of Paris and a member of the Academy of Sciences, he was already famous and respected in the intellectual world. He had published over three hundred scientific papers, books, and articles in physics and philosophy as well as mathematics. One of the last scientific "universalists," he was able to

Plate 1.1. Henri Poincaré. (Courtesy of Mittag Leffler Institute)

grasp and contribute fundamental ideas in several different fields. His influence in mathematics and science was enormous.

Poincaré was enjoying his walk. He tried to completely empty his mind of mathematics. His steps had taken him toward the Seine. As the Eiffel Tower came into view, he remembered its inauguration at the Paris Exposition of 1889. That year had been as important in his career as it was in Gustave Eiffel's, for it was then that Poincaré had won the prize established by King Oscar II of Sweden and Norway for the work that is the subject of this chapter: his paper on the dynamics of the three-body problem.

For the exposition, Eiffel had built the tallest structure in the world. After more than a century, it has become a symbol for Paris throughout the world. Parisians did not like it at first, and many still dislike it today. This unearthly creature appears to dominate so much that is of more intimate beauty and architectural value. A tourist visiting the capital often goes first to the Eiffel Tower. L'Île de la Cité with the Notre Dame cathedral, the Champs Élysées guarded by the Arc de Triomphe and the Concorde square, the Opéra, the Tuileries Garden and the Louvre Museum, Montmartre and the Latin quarter: all are overshadowed by the great tower. It stands in its effrontery, straddling the crowds on its four huge legs.

Eiffel's vision and daring were admirable, but Poincaré could not admit approval of the tower. Too well established and respected as a mathematician, as a professor, and as a citizen, he would not take the risk of espousing a controversial view and perhaps appearing ridiculous in the eyes of his fellow countrymen. For a mathematician, there is little danger of this within his own profession. No matter how surprising the final result, as long as one shows that each statement follows logical reasoning and computation, one seldom runs the risk of being considered ludicrous. In the fine arts, architecture, or literature, where judgment is subjective and uncontrollable criteria are important, a swing of the public mood can destroy a career. Often it is just a question of luck. Sometimes it is a matter of seizing—or mistaking—the social and historical moment. Such unfortunate events rarely occur in mathematics, although timing, accidents, and the public mood do exert a more subtle influence.

Poincaré was one of very few among his peers who knew and could follow almost every achievement in mathematics up to their time. The explosion of research and information has made this impossible today: mathematicians, like their fellow scientists, are each confined in limited worlds, largely ignorant of progress and issues outside. Today, at the major quadrennial meeting, the International Congress of Mathematicians, participants cannot always understand even the *titles* of papers outside their specialties.

As he walked, Poincaré found himself recalling earlier thinkers and the tricks that fate had played on them. Gauss, the most famous mathematician in the first half of the nineteenth century, refrained from publishing his discovery of *non-Euclidean geometry* for fear of the "screams" of his con-temporaries. Gauss knew that they would fail to understand his abandon-ment of Euclid's "common sense" axioms. Some years later, Janos Bolyai, a young Transylvanian officer whose father had also worked on the prob-lem, developed similar ideas and *did* publish them. Bolyai even wrote to Gauss on the subject. In his reply, the German mathematician revealed his earlier thoughts. Yet today it is Gauss who is much more widely cited. Ironically, there is little recognition of Bolyai's work, which became known only several decades after his death. Other instances occurred to Poincaré: Galois, who died at the age of twenty-one in a duel, left behind a short paper, to be appreciated by the French Academy only half a century later. Today, *Galois theory* is a mathematical field in itself.

As he strolled home through the streets of Paris, Poincaré's thoughts returned to his own problem. Was he being too conservative in his ap-proach to it? No, he must proceed in a logical way. His mind drifted back to the scene around him. Just as it seemed that he had achieved nothing with his walk, revelation came like a flash of light. *Now* he understood. He was given a sudden, graphic hint of the consequences of his finding, eight years earlier, that certain motions of the restricted three-body prob-lem are unstable. His reasoning and computations had been entirely cor-rect. It was their implications that he had not been able to accept. He now saw that the problem, which he wanted to present in the book that would summarize half his research career, did indeed exhibit unexpected and strange behavior. He had no name for it and dared not pursue it further for the moment, but he had glimpsed a new land. Although it appeared incredible to his clear and straightforward beliefs, he would have to come to terms with it.

We shall spend the rest of this chapter describing the background to this startling insight. This will require a trip through several areas of mathemat-ics and science, for one hundred years later we have found more than a mere name for the bizarre behavior that Poincaré glimpsed that day in Paris. It has become a new way of thinking. We call it *chaos*.

Newton's Insight

What happened in Poincaré's mind at that moment? To appreciate it, we must go back more than two centuries, to the midsummer of 1687, when Sir Isaac Newton published his masterpiece, *Philosophiae Naturalis Prinicipia Mathematica* (The Mathematical Prin-ciples of Natural Philosophy). The main goal of this long trip back in time is to trace a crucial notion in the development of mathematics and physics: that of a *differential equation*. The rudiments of differential equations were already known at the end of the sixteenth century to the Scottish mathematician John Napier, the inventor of logarithms. (The name derives from "na pier" [no peer], meaning a *free man*.) But it was Newton who raised differential equations to their present central position in science. He showed not only how to express problems in physics using them, but also developed the basic mathematical tools needed for their solution.

The style of the *Principia* is difficult. The ideas Newton proposed are hard to understand if one has no prior feeling for them. Newton even claimed that he wrote the book in this manner with a purpose: "to avoid being baited by little smatterers in mathematics." His major physical contribution is the idea of connecting gravitation with the dynamical behavior of the solar system: its evolution in time. Prior to Newton's work, it was widely believed that gravitation acted only on bodies close to the earth's surface. By proposing that its force extends throughout the universe, Newton realized that the moon's motion, the tidal effects, and the precession of equinoxes could all be explained by gravitation. The conclusions drawn in his book brought Newton an immediate scientific reputation.

Although the *Principia's* impact in physics was to be enormous, the longer-term consequences, which make the work celebrated even three centuries after its first publication, lie in its mathematical contributions. The heart of this is the *differential* and *integral calculus*, topics which form the course taken by many of today's university freshmen (a course loathed by some), and on which practically all modern science and engineering are founded. It is, however, hard to see in *Principia* a resemblance to a modern calculus textbook. Newton's notation appears old-fashioned to a modern reader. His geometric language is hard for contemporary scientists to follow. The conventional formalism is closer to that proposed by Gottfried Wilhelm von Leibniz, who is considered the co-creator of calculus (and with whom Newton had bitter disputes about scientific priority). Ironically, the new approach to differential equations due to Poincaré, which we shall describe below, is in some ways closer in spirit to Newton's, for it is geometrical in nature. Many mathematicians "think in pictures," even though they may eschew them in their published work. Such an approach might provide a more attractive entrée for the student, but has yet to find its way into freshman textbooks.

More broadly, Newton showed how to derive mathematical models that would describe many other physical processes. In addition to celestial mechanics—the study of the movements of heavenly bodies—he laid the foundations for rational mechanics in general, creating the theory necessary to explain and unify the observations of Kepler, Galileo, and others. Like the axiomatic approach of Euclid to geometry, rational mechanics attempts to explain a mass of disparate phenomena in terms of a few basic laws. Calculus was central to this approach and Newton used it to bring into being and analyze the mathematical objects we now call *differential equations*. What is a differential equation and why is it so important? For the next few pages we will leave Poincaré and Newton to explore this notion.

A LANGUAGE FOR THE LAWS OF NATURE

Mathematics provides a particularly useful language in which to express physical laws such as those that Newton proposed should govern the motions of gravitating bodies. However, one must be careful. Over the centuries, mathematicians have appropriated many common words of English as names for their own special concepts and constructions. This makes our task difficult: sometimes we intend the usual sense of the word, with all its associated implications, and at others we require only the technical, "mathematical" meaning. In this book we shall generally signal such a technical usage by *italicizing* the word or phrase in question.

One of the most useful mathematical "sentences" is a *differential equation*. This is a device relating the rates of change of *variables* that describe the state of a (physical) system to the current values of those same variables. To illustrate, we will use a simple example: a falling ball. Suppose we drop a ball from a high window so that it falls straight down. At any moment during the fall its *state* can be described by two *variables*: its position or height, h, and its velocity, v. These quantities are called variables because they change as time progresses and the ball falls. They are interpreted as *functions of time* that take the values $h(t)$ and $v(t)$. The parenthetical t refers to the fact that they depend upon time, which is conventionally denoted by this letter. Graphs describing the monthly changes in unemployment or interest rates represent examples of such functions of time. The basic premise in our case is that the functions h and v exist and can be found by *solving* an appropriate differential equation, as we shall describe below.

Both of these variables are necessary to determine the state of the ball. Thanks to Newtonian mechanics, we know that the rate of change of the ball's momentum is equal to the forces on it due to gravity, air resistance, the wind, and anything it strikes on the way down. Momentum is the product of mass and velocity, but here the mass doesn't change, so we can say that the *rate of change of velocity*—called *acceleration*—multiplied by the mass is equal to the forces acting on the ball. (This is the basis for the famous equation $F = ma$ of freshman physics.) In turn, *velocity* is just the *rate of change of position*. Supposing the ball is sufficiently heavy that the effects of the air may be ignored, and supposing it has a free path, only the gravitational force remains. Newton's gravitational theory tells us that, near the earth's surface, this force is equal to the product of the mass of the ball, m, with a constant, conventionally called g. Thus we have the relationships:

rate of change of position = velocity
rate of change of momentum = gravitational force,

or, in the shorthand notation invented by Leibniz, which we use today:

$$\frac{d}{dt} h = v$$

$$\frac{d}{dt} (mv) = -mg.$$

The first equation is simply the definition of velocity, the second is the statement of one of Newton's laws of motion. The notation d/dt means "rate of change." More literally, in the first equation, for example, $(d/dt)h$ denotes the ratio of incremental changes (dh) in the variable h to those in t, time. (Newton's different usage of a dot over the quantity—$\dot{h}$, $\dot{v}$—to denote rate of change is still used in physics and engineering.) The negative sign in the second equation means that gravity acts downward, while we conventionally measure height and velocity upward. Since the mass is constant and does not vary with time, it could actually be canceled out of the second equation. As we shall see in a moment, these two relationships will uniquely determine the functions h and v, once we have specified the values $h(t_0)$ and $v(t_0)$ of both variables at the instant $t = t_0$ when the ball is released.

This is the first and last differential equation we will write down as such, for, thanks to the approach pioneered by Poincaré, we can give a beautiful geometrical characterization of the equation and, in so doing, avoid formulas altogether. Watching a falling ball, we judge its height and velocity at some instant and can perhaps predict when it will strike the ground. Position, speed, and time are all mixed up in our perception of the scene. To disentangle them, we move from physical space, in which the ball falls, to a mathematical *phase space* or *state space*, in which we represent the differential equation describing its motion. Now phase space is not space in the same sense as the three-dimensional world we inhabit: it merely takes its cue from physical space to represent the quantities describing the state of the ball in pictorial form. As we saw above, the state is characterized by position h, and velocity v, both of which depend upon time. Phase space separates these quantities by plotting the path followed by the ball as a curve describing its position and velocity at each instant.

In figure 1.1a the horizontal axis indicates height and the vertical axis, velocity. Here phase space is the plane of the page. Positive velocities correspond to the ball rising (having been thrown up), and negative velocities indicate the ball is falling earthward. The arrows show the direction of increasing time. Suppose the ball is dropped from rest at a height of 20 meters (the point a_0). It follows the curve a_0, a_1 to strike the ground (0 meters) at a velocity of -19.81 m/sec. Dropped from 25 meters, it goes from b_0 to b_1 to land at -22.15 m/sec. Thrown upward from the ground at the same speed ($+22.15$ m/sec), it starts at b_2 and follows b_2, b_0, b_1, first rising to 25 meters and then descending, following exactly the same path, in phase space, as the ball dropped from that height. Figure 1.1a displays, in one picture, *all* possible motions of the ball as a *family* of curves. The word "family" indicates the fact that the curves all derive from the same differential equation.

These curves in phase space are parabolas, one of the conic sections of Greek mathematics, illustrated below in figure 1.5d. However, they must not be confused with the parabolic paths that balls thrown at an angle actually describe in physical space: all the motions in our phase space correspond to balls falling vertically or thrown vertically up and retracing the same path in physical space on the way down; see figure 1.1b. We say that the system has *one degree of freedom*, since there is only one direction in which the ball can travel, and only one ball. Each degree of freedom implies two dimensions in phase space: one for position and one for velocity. To represent the motion of a ball thrown at an angle, we would need a phase space having four dimensions: two for position, vertical and horizontal, and two for the corresponding components of velocity, the vertical and horizontal speeds, since the ball can now travel along as it travels up or down. In this case we say there are *two degrees of freedom*. Two balls would require further doubling the dimension of phase space, for we would need two position and two velocity variables for each ball. In general, both position and velocity are *vectors*, each with n *components*, each component being a function of time specifying the relevant quantity in a given direction. We cannot (easily) draw the associated four or higher dimensional pictures, but we can deduce some of their properties.

We pause to remark that this differential equation is the simplest possible model for an object thrown or falling vertically. So Newton's laws successfully and economically explained the famous (and probably apocryphal) experiment in which Galileo simultaneously dropped balls of various masses from the leaning tower of Pisa, finding that they all reached the ground at the same time. Recall that the mass canceled out in the second equation above: a falling object's acceleration is not affected by its mass. But the model does not include "second order" effects such

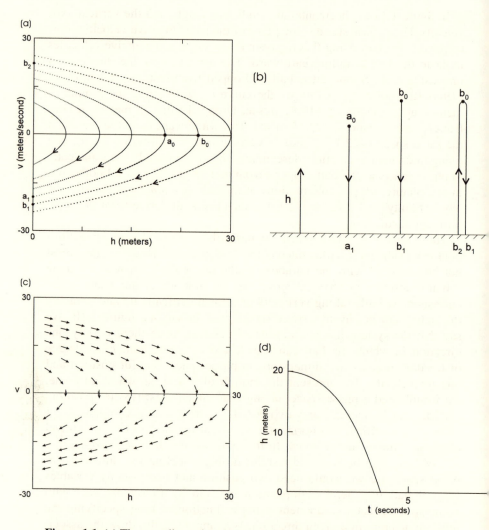

Figure 1.1. (a) The two-dimensional phase space of the falling ball; (b) some corresponding motions in physical space; (c) the vector field; (d) position $h(t)$ as a function of time.

as air resistance. Like all theories and models, it provides an idealized and simplified view of reality. Physics is successful largely because scientists have learned how to ignore irrelevant information and concentrate on the essence. Additional effects can be added to the model if they seem necessary. To accurately predict the motion of a feather or a falling leaf, we would have to include forces due to the air through which it floats.

MODELS OF REALITY

Models play a central role in science. In studying the complexities of genetics, biologists resort to breeding generations of fruit flies and studying their mutations. The fruit fly serves as a model for other, higher organisms. One hopes that it captures key aspects, so that its relative simplicity will permit a better understanding of genetics in general. In the same way, the differential equations of Newtonian mechanics provide *models* for physical phenomena like falling balls or, more grandly, the motion of the planets. Such models provide *approximations* to the real phenomena. Elasticity theory can predict the stress in a bridge truss to a few percent, but not all models are this accurate: the differential equations modeling neural networks offer rather a metaphor for the real system. Metaphors suggest literature, and this is an appropiate comparison. Much as the poet W. H. Auden has spoken of the "secondary worlds" of art, music, and literature, differential equations inhabit a secondary mathematical world, through which we hope to better understand our own "real" world. As such, they are subject to all the rules of the pure mathematical world, and all its tools and techniques are available for their solution. We shall now continue to explore some of these.

How were the curves of figure 1.1 obtained? By *solving* or *integrating* the differential equation that we wrote earlier in verbal and symbolic form. As we have already stated, to do this Newton and Leibniz invented the integral calculus. Integration turns the equations relating rates of change into formulas expressing the positions and velocities of the ball at every instant of time: it reveals the functions h and v which we described earlier. Roughly speaking, one sums up (integrates) all the infinitesimal changes occurring as time passes to find the values $h(t)$ and $v(t)$ after a finite time has elapsed. We will describe the process from our geometric viewpoint.

The rate of change expression effectively attaches an arrow to every point in the phase space, indicating (by its length) the rate and (by its angle) the direction with which the changes in position and velocity are taking place there. Together, all these arrows, called vectors, form a *vector field*; see figure 1.1c. To solve the differential equation, we start at a point such as a_0 and follow the arrows, producing a curve that is at every point *tangent* to the vector field defined by the differential equation. Standing on the plane of phase space, one can imagine taking a step in the direction of the vector at one's feet, of distance proportional to its length. One then repeats the process for the vector based at the landing point, and so on. Letting the steps get smaller and allowing their number to increase, the polygonal path so produced approaches a smooth curve.

Thus, in a manner somewhat like that of the wind, which leaves a memory of its track in the waving ears of a wheatfield, or a river current that carries flecks of foam and driftwood on its surface, leaving local evidence of its passage, the vector field drives the individual solutions of the differential equation, starting at each point in phase space, to form a global *phase portrait*. The complete picture that results is the *general solution* or *flow* of the differential equation. The paths threading through the picture are called *solution curves, orbits* or *trajectories*. The family of parabolas in figure 1.1a provides one example of a general solution. We will meet many more. Finally, the path in phase space can be replotted as a function of time: a graph showing the height of the ball at each instant of time; see figure 1.1d.

Newton considered his discovery so important that he published it as an anagram, which, if decoded, stated: *Data aequatione quotcumque fluentes quantitae involvente fluctiones invenire et vice versa*, or: *Given equations involving how many soever flowing quantities, the flow can be determined, and conversely*. This expresses the connection between the differential equations and the geometric image of the flow. The word flow recalls Newton's term *fluent* for a dynamically varying quantity; the rates of change, now called *derivatives*, he named *fluxions*.

To solve a differential equation means to find its general solution: its flow. Sometimes we require only a single curve starting at a particular point, such as a_0 in figure 1.1. Then we speak of solving an *initial value problem* for the differential equation. This is the approach of the calculus textbook: the answer that the author expects from the student is a formula telling the values of the state variables, such as position and velocity, at each moment in time, given a particular initial position and velocity. Unfortunately, the class of equations that can be explicitly solved, and such formulas obtained, is very small. (On this point most calculus textbooks are silent: people do not like to be told that they have worked hard to learn a method that is not very useful!) By contrast, in the geometrical approach of Poincaré, we attempt to find qualitative properties of the solution, *without asking what the formula for the solution is*. These qualitative methods apply to a much larger class of differential equations.

But first we must ensure that a solution exists: that the variables of the problem can indeed be represented as functions of time, even if we cannot find explicit formulas for them. In some cases there may be no smooth curve having at every point a tangent vector belonging to the given vector field. The vectors might "contradict" each other, so that one cannot follow the arrows in a continuous manner. Under such circumstances the *existence property* is violated and it makes no sense to study nonexistent objects. It can also happen that through a given point we can draw two or more smooth curves having tangent vectors at every point, vectors taken again only from the vector field specified by the differential equation (see the curves $r_{-1} \, r_0 \, r_1$ and $s_{-1} \, s_0 \, s_1$ in fig. 1.2). In this case we say that the *uniqueness property* is not satisfied.

Questions such as *existence* and *uniqueness* are fundamental to the theory of differential equations. A large class of vector fields does satisfy both the existence and uniqueness properties, and this includes almost all the ones that model processes in physics. This fortunate fact lends physics its spectacular powers of prediction. Let us imagine for a moment what might happen if the uniqueness property were not fulfilled by an initial value problem modeling a process in a nuclear reactor. Then, for a given input, one could not decide which reaction takes place: a safe one, leading for example to equilibration at the desired operating condition, or a dangerous one, which might cause a nuclear catastrophe. In our more modest example of the falling ball, a model incapable of predicting the ball's path or time of flight from its initial position and velocity would be of little use.

Things are, of course, not as simple as they may at first appear. Differential equations provide models in fields as diverse as physics, chemistry, biology, sociology, economics, and psychology. Their behaviors can differ wildly. Even the basic questions of existence and uniqueness, though easy for some classes of equations, remain unsettled for others. For example, the *Navier-Stokes equations*, which model the motion of an incompressible fluid such as water in three-dimensional physical space, are not yet proven to satisfy the *global existence* property. In spite of having discovered many features of their *possible* solutions, we still do not know if these solutions exist in full generality.

There is some controversy among mathematicians and physicists on such matters. The latter typically do not worry much about existence questions. For them an equation is primarily a means to obtain those properties of the solutions needed to understand and predict physical phenomena. This is perfectly reasonable and natural. But mathematicians are, first of all, concerned with existence and uniqueness, and only go on to study the properties of solutions after they have settled these questions. Their reasons are also clear. They are primarily interested in mathematical objects, and are understandably wary of things that may not exist. (In the world of mathematical logic, *any* statement concerning such objects is true, since it is predicated on something that does not exist.) Ignoring these fundamental matters, one might well draw erroneous physical conclusions. However, for cases like the Navier-Stokes equations, which form a well-established model, specialists have few doubts that solutions do exist, and most believe that some day the existence question will be settled. In doing so, subtle properties of solutions with as yet unknown physical implications may emerge. Until then, it seems reasonable to *assume* that solutions exist and to spend time studying their properties relevant for applications.

These kinds of questions demand contributions from both mathematicians and physicists. They are central to *applied mathematics*, a domain with an ill-defined boundary, lying between *pure mathematics*—the unearthly realm of axioms and theorems—and the real, messy world we all inhabit. Applied mathematicians are largely concerned with the construction and analysis of mathematical models. The best work in their field moves effortlessly from practical applications to purely mathematical questions and back: a true "technology transfer" between the real and secondary worlds.

We mentioned the notion of *global existence*. To understand this, recall that each of the solution curves of figure 1.1a is created by a point representing the system's state, moving in time. Global existence implies that the curve is defined for every moment in the past, present, and future. This does not necessarily mean the trajectory comes from infinity and returns to infinity in the phase space. It could be a closed loop, around which the state travels infinitely often, repeating its path over and over again, as does the loop q_0, q_1, q_2 in figure 1.2. This is an example of a *periodic orbit*. Perhaps the most familiar instances of such motions are the annual cycle of the earth around the sun, and the monthly orbit of the moon about the earth.

In contrast to the notion of global existence is that of *local existence*. This means only that the curve exists for some short time interval about the initial instant (for example, the small segment $p_{-1} p_1$ around p_0 in fig. 1.2 is traversed by the solution in such a short time interval). The differential equations of celestial mechanics that we shall deal with generally satisfy the hypotheses of local existence, but, as we shall see in chapter 3, global existence does not always hold for them. Global existence implies local existence, but the converse is not necessarily true.

Another important property we rely on is the *continuity of solutions with respect to initial data*. Roughly speaking, this means that trajectories having close initial conditions (data), remain close at least for a while; see the curves $c_0 c_1$ and $d_0 d_1$ in figure 1.2, with close initial conditions c_0 and d_0. This property also commonly holds for differential equations. However, if we ask that globally defined solutions, with close initial conditions, stay close together for *all* time, then we are demanding a rather rare property, called *stability*; see the curves $e_0 e_1$ and $f_0 f_1$ in figure 1.2. Stability is also a fundamental notion in the study of differential equations. It arose from a traditional problem in celestial mechanics—the stability of the solar system—and was systematically developed in the work of Aleksandr Liapunov at the end of the last century. We will return in chapter 4 to discuss it in more detail.

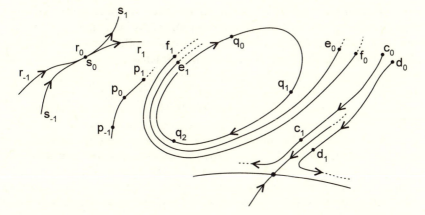

Figure 1.2. Another two-dimensional phase space, illustrating failure of uniqueness, local and global existence, continuous dependence on initial conditions, and stability. See text for explanation.

Almost all of our examples and explanations thus far have been couched in a two-dimensional phase space: the plane. If we were limited to two state variables for each problem we would not get very far. Fortunately, differential equations can also be defined in more general, "abstract" spaces. The extension applies not only to dimensions higher than two; it can be pushed further to include abstract geometrical objects called *differentiable manifolds*. The first natural extension is to three-dimensional space. The definitions made and conclusions drawn about existence, uniqueness, and continuous dependence of solutions of differential equations in the plane can all be extended. We can generalize to higher-dimensional spaces, having dimensions four, five, or even 7,058. It does not matter how high we go as long as the dimension remains finite. Such spaces are beyond our physical imagination. We cannot envision even four- or five-dimensional objects, but through analogies with dimensions one, two, and three, we can draw useful conclusions about them.

The study of such high-dimensional phase spaces is more than a game invented by mathematicians. It may seem paradoxical, but most dynamical problems arising in physics and engineering are easier to tackle in phase spaces with more than three dimensions or on differentiable manifolds. We have already pointed out that four dimensions—two position and two velocity components—are needed to describe the state of a ball thrown at an arbitrary angle. To describe the state of a stick tumbling about in the air after being thrown, we would also need angular variables to describe its orientation, and the associated angular velocities. The simplest possible model for an automobile requires fifteen or so positions and fifteen velocities to describe the relative motions of the body, wheels, and suspension components: a phase space of dimension thirty. As ball, stick, and car follow their paths through our three-dimensional world, their mathematical models evolve in state spaces of unearthly dimension.

MANIFOLD WORLDS

What is a *differentiable manifold*? Let us start again in two dimensions. A two-dimensional differentiable manifold is a geometrical object that can be locally approximated by part of a plane, like a papier-mâché mask, in which small flat pieces of paper overlap to create a smoothly curved surface. The surface of a sphere is an example of such a two-dimensional manifold: around every point one can find a small neighborhood that can be well approximated by a planar region. (To a mathematician, the *sphere* is the two-dimensional surface alone, excluding the solid ball inside.) Although we live on a globe, we typically perceive it as a planar surface, and our local road maps rely on this convenient approximation. Indeed, it took humans thousands of years to realize that the earth is not a plane. Another classical and important example of a two-dimensional manifold is the *torus*, a mathematical name for the surface of a doughnut. The region around any point on the torus can also be approximated by a small piece of plane. Locally, the torus, the sphere, and the plane are indistinguishable: we have to "get outside of them" to appreciate their curvature and global structure. (Imagine how an alien civilization living on a toroidal planet might discover that fact.) Figure 1.3 shows these three common examples; can you think of any others? The term "differentiable" implies that the manifold in question is smooth and therefore (differential) calculus can be carried out on it.

In the same way, the notion of a *k*-dimensional manifold may be defined for any number *k* larger than two. If *k* equals three, for example, one requires that a small piece around any point of the manifold be like a small piece of our usual three-dimensional space. Note that, in general, a manifold cannot be entirely or *globally* approximated by the corresponding *flat* space on which it is modeled. A sphere cannot be unfolded onto a plane and

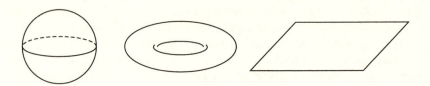

Figure 1.3. Sphere, torus, and plane.

it really is different from the latter: for example, it has no boundaries—no edges—but is finite nonetheless. The straight lines and planes of high school geometry and the ambient space of our everyday experience are examples of flat or *Euclidean* spaces, of dimensions one, two, and three, respectively.

We can define a differential equation on a differentiable manifold such as a sphere. Then the vector field will be confined to it. This means that at every point on the sphere a vector is attached, contained in the plane tangent to the sphere at that point. To solve an initial value problem on the sphere means to find the smooth curve on it having at every point a tangent vector belonging to the vector field (see fig. 1.4). We shall discuss vector fields on tori in chapter 5.

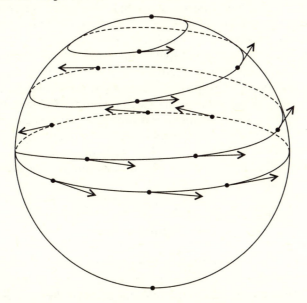

Figure 1.4. A vector field on a sphere and some of its solution curves.

We have already seen that multidimensional phase spaces may be needed to describe physical situations, but why do we need the apparent extra complication of differentiable manifolds? A major reason is that their use can actually make differential equations easier to solve by *reducing* the dimension of their phase spaces. Newtonian mechanics reveals a host of *conserved quantities* or *constants of motion*: properties of dynamical systems that might be thought typically to change, but which actually retain their initial values for all time as the system evolves. In doing so they impose fixed (algebraic) relationships among the state variables, confining them to regions of phase space delineated by those algebraic equations. These regions are usually differentiable manifolds.

The *angular momentum* of a rigid body, such as a stick tumbling in space, provides a good example. It is a consequence of Newton's laws of motion that, if no turning forces (*torques*) act on a body, its angular momentum will retain its magnitude unchanged. The effect is used by a quarterback, who puts a spin on the ball to stabilize it as he throws a pass. We need three components, say a_1, a_2, and a_3, to specify the angular momentum of a body in three-dimensional (physical) space. The constraint of constant magnitude implies that $a_1^2 + a_2^2 + a_3^2 =$ constant, which is exactly the equation defining a two-dimensional sphere in three-dimensional space. The equations for the rigid body can be solved in this reduced phase space rather than in the larger, three-dimensional state space. In chapter 5 we shall describe this reduction method in some detail. And even when such a reduction is not possible for all solutions, certain special sets of orbits may still live on smooth manifolds of lower dimension, which can then serve as a "skeleton" in our attempts to construct the full phase portrait.

The study of manifolds forms an important part of *topology*, the area of mathematics concerned with the intrinsic properties of objects that persist under distortions of size or shape. It is no accident that Poincaré was one of the inventors of topology, or "analysis situs"—the study of position, as it was then called. Now that we have described some notions in this and in the theory of differential equations, let us continue our attempt to understand his thoughts.

THE *n*-BODY PROBLEM

What was the problem that had occupied Poincaré for so long? He was working on one of the oldest dreams of humankind: to understand the motion of celestial bodies—the sun, planets, and moons of our solar system. The first complete mathematical formulation of this problem had appeared in Newton's *Principia*. Since gravity was responsible for the motion of planets and stars, Newton had to express gravitational interactions in terms of differential equations.

A fact, which Newton at first simply assumed true, is that most celestial bodies can be modeled as point masses. Consider first the case of two bodies. Newton's gravitational law states that the force acting on each body is directly proportional to the product of the masses, inversely proportional to the square of the distance between their centers, and acts along the straight line joining them. Thus gravity acts upon the bodies, considered as solid spheres, in exactly the same way that it would act on point particles having the same masses as the spheres and located at their centers. Newton developed his whole theory on this assumption and, just before publication, realized that it needed justification. It took him some time to find a rigorous mathematical proof of the theorem, so publication of the *Principia* was delayed. However, this fundamental result permitted the formulation of the famous *n-body problem*, one of the most important questions in the historical development of mathematics as well as physics.

The physical problem may be informally stated as: Given only the present positions and velocities of a group of celestial bodies, predict their motions for all future time and deduce them for all past time. More precisely, consider n point masses m_1, m_2, ..., m_n in three-dimensional (physical) space. Suppose that the force of attraction experienced between each pair of particles is Newtonian. Then, if initial positions in space and initial velocities are specified for every particle at some "present" instant t_0, determine the position of each particle at every future (or past) moment of time. In mathematical terms, this means to find a global solution of the initial value problem for the differential equations describing the n-body problem. For n equal to two this is called the two-body problem, or the *Kepler problem*, in honor of Johannes Kepler, the German astronomer who, with his laws interpreting the astronomical observations of Tycho Brahe, provided Newton with inspiration for the gravitational model. (See fig. 1.5.)

The differential equations of the two-body problem are easy to solve. One can prove that the path followed by one particle with respect to the other always lies along a *conic section*. (Conic sections derive their name from the fact that they can be obtained by slicing a cone at different angles, as in figure 1.5c). This means that the orbit described in physical space may be a circle, an ellipse, a parabola, a branch of a hyperbola, or a straight line.

At first sight the equations might seem forbidding: a complete description of each particle's state requires three position variables and three velocity components, for both particles move in three-dimensional physical space. Thus the phase space has twice six, that is, twelve dimensions. We say that the system has six *degrees of freedom*: one for each of the position variables needed to describe it. How can we analyze solutions that wind about in a twelve-dimensional space?

Here the various conservation laws of Newtonian mechanics come to our aid. First, since forces act only along the line joining the particles, once set in motion in a plane, the particles remain forever in that same plane. Thus we only need two positions and two velocities each, or eight variables in all. In addition, *linear momentum*—the sum of the products of the masses and their velocities—is conserved, so the position of the center of mass does not change with respect to the bodies (it lies somewhere on the line connecting the particles, nearer to the heavier body). Thus we need only specify the *relative* position of one particle with respect to the other,

(a)

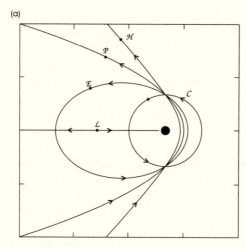

(b)

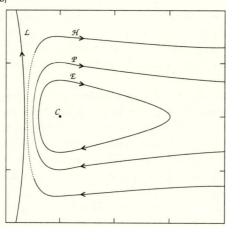

Figure 1.5 The two-body
problem: (a) several possible
motions of a gravitating particle with respect to another particle in physical space, and (b)
the same motions represented
in phase space. (c) Conic sections are obtained by slicing a
cone at different angles. Here
C denotes a circle, E an ellipse,
P a parabola, H a hyperbola,
and L a line segment.

(c)

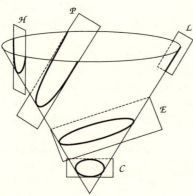

and we are down to two position variables and two velocities, or two degrees of freedom. Finally, the fact that no torques act means that angular momentum is also conserved. This is the effect that causes a ballet dancer or an ice skater to spin faster when she draws in her arms: the rotational and radial motions are related and each determines the other. So, the equations can be reduced to just two, describing the rates of change of the distance between the particles and their relative velocity along the line joining them. Only a single degree of freedom remains.

The complete solution of the Kepler problem was first given by the Swiss mathematician Johann Bernoulli in 1710. The name Bernoulli, which originated in Antwerp, appears frequently in mathematics and physics. At least eight members of the Bernoulli family, belonging to three generations, made important scientific contributions. Johann was born in Basel in 1667 and became a professor at the university in his hometown in 1705, succeeding his older brother Jakob in this position. Daniel, one of Johann's sons, is known for his contributions to fluid mechanics and the kinetic theory of gases.

Unfortunately, the two-body problem has proved to be the only easy one among all the *n*-body problems. For *n* larger than two, in spite of an enormous expenditure of effort, no other case has been solved completely. The problem even became a topic of conversation in the fashionable salons of Paris, exciting the interest of such people as Voltaire and Madame de Maintenon. Most of the great mathematicians of the eighteenth and nineteenth centuries tackled the governing differential equations but were unable to make much progress, although they contributed much to mathematics and the theory of differential equations in the course of their attempts. We will later have the opportunity to examine some of their achievements.

KING OSCAR'S PRIZE

"Given a system of arbitrarily many mass points that attract each other according to Newton's laws, try to find, under the assumption that no two points ever collide, a representation of the coordinates of each point as a series in a variable that is some known function of time and for all of whose values the series converges uniformly.

This problem, whose solution would considerably extend our understanding of the solar system, would seem capable of solution using analytic methods presently at our disposal; we can at least suppose as much, since Lejeune Dirichlet communicated shortly before his death to a geometer of his acquaintance [Leopold Kronecker], that he had discovered a method for integrating the differential equations of Mechanics, and that by applying this method, he had suc-

Plate 1.2. Gösta Mittag-Leffler. (Courtesy of Mittag-Leffler Institute)

ceeded in demonstrating the stability of our planetary system in an absolutely rigorous manner. Unfortunately, we know nothing about this method, except that the theory of small oscillations would appear to have served as his point of departure for this discovery. We can nevertheless suppose, almost with certainty, that this method was based not on long and complicated calculations, but on the development of a fundamental and simple idea that one could reasonably hope to recover through preserving and penetrating research.

In the event that this problem nevertheless remains unsolved at the close of the contest, the prize may also be awarded for a work in which some other problem of Mechanics is treated in the manner indicated and solved completely.

Thus read the announcement in *Acta Mathematica*, vol. 7, of 1885–86. Gösta Mittag-Leffler, a prominent mathematician in Stockholm and editor in chief of the recently founded journal, had taken a bold step. He had

Plate 1.3. King Oscar II of Sweden and Norway. (Courtesy of Uppsala University Library)

convinced King Oscar II of Sweden and Norway to establish a substantial prize and medal to be awarded to the first person who obtained the global general solution of the n-body problem, and to mark, through its award, the king's sixtieth birthday on 21 January 1889.

After some debate regarding the difficulty of getting the world's most prominent mathematicians to agree to serve together, an international jury

Plate 1.4. Karl Weierstrass. (Courtesy of Mittag-Leffler Institute)

was formed of Karl Weierstrass of Berlin (Mittag-Leffler's former teacher), the French mathematician Charles Hermite, and Mittag-Leffler himself. This jury set four problems, of which only the first, cited above, concerns us, although it is interesting to note that the fourth related to a group of "Fuchsian" functions on which Poincaré had also worked.

We must take a moment to explain some technical terms, which will recur several times in the book. In a *series expansion* or *approximation*, one seeks the solution (of the differential equation) as an infinite sum of *terms*, each of which solves a simpler, subsidiary equation. The first term provides a rough approximation, the sum of the first and second, a better approximation, and so forth. If the corrections decrease to zero and the sum of all terms is finite, we say that the series *converges*. In the present context, each term is moreover a function of time, and if convergence occurs for all times future and past, the series *converges uniformly*. The *theory of small oscillations* refers to the equations which generate the first-order approximations, the so-called linearized equations. The use of series methods in celestial mechanics was pioneered by Laplace and Lagrange, whom we shall meet in chapter 4, where we will also discuss linearization.

The closing date for submissions was set as 1 June 1888. The financial reward of 2,500 crowns was not outstanding (it corresponded to about a third of Mittag-Leffler's annual salary), but the prestige of such an award was, at that time, equivalent to a Nobel Prize today. Immediately following the announcement, the heavy artillery of the mathematical world was directed on the *n*-body problem. Many people wanted to solve it, but in the end only five of the twelve entries to the competition attempted it.

POINCARÉ'S ACHIEVEMENT

Henri Poincaré was just thirty-one years old when he heard about King Oscar's prize. Though he was already becoming established and internationally known, he realized that if he were to win the prize it would further his career enormously. The problem was extremely difficult, no doubt, but it might be worth a try. Such opportunities are rare, and it might be a long time before another would come.

It was a rainy night and he felt tired. It was time for sleep. Psychologically unprepared for the news he had just received, he could not decide whether to attempt the problem or not. Five years before, he had failed in a similar competition. The essay he submitted for that Grand Prix had not been considered good enough. This time things were different. He felt stronger and better prepared; still, a tinge of anxiety remained.

There is always risk involved in tackling a hard question. One must bring knowledge and skill and new ideas to it, and be prepared for a tremendous amount of work. It is not enough to be clever and able to create new ideas, to be endowed with good intuition and with analytic and synthetic powers of thought; there is also an important element of luck. Above all, one must be strong and stubborn. After a night of bitter disappointment, when it seems that months of work are rendered useless because of a simple technicality overlooked at the outset, one still must get up next morning full of energy and start again with even greater enthusiasm. It is so much easier to stick to familiar territory, to perform only in areas in which one is well rehearsed. Few have the courage and endurance to explore the unknown.

In bed but unable to sleep, Poincaré thought about his life, about happiness, about success. He trusted his own powers of understanding, but fatigue dulled his customary optimism. Subsequently he would write: "Life is only a short episode between two eternities of death, and even in this episode, conscious thought has lasted and will last only a moment. Thought is only a gleam in the midst of a long night. But the gleam is everything." He had still not decided whether to take up the challenge when finally he fell asleep.

On waking the next morning, the decision had been made. He would begin. It was the only possible choice. He knew he had to do it.

Celestial mechanics had long been a favorite subject of Poincaré's. As a child he had enjoyed surveying the night sky from his family's backyard. He had learned the names of the constellations and been delighted by the wandering of the planets. It was so quiet outside. The cool and the fresh aroma of summer flowers was pleasant. He felt small but secure at the same time, a part of something far greater. Everyday problems seemed unimportant: the stars had been there long before he was born, and regardless of what he might do, they would long remain.

Poincaré's first mathematical paper in the field of celestial mechanics, published in 1883, treated some special solutions of the three-body problem. A second paper followed a year later. In spite of their depth and importance, these contributions could not be considered breakthroughs. The prize question would require much more, but Poincaré's previous work had given him some insight into the subject. He had always wanted to study it further, yet other projects had taken priority, and he had never found sufficient time to return to it. In the meantime he had developed new approaches to differential equations which might be useful for this problem. He felt he had good intuition about the right line of attack.

Nevertheless, the beginning was difficult. Poincaré quickly reached the limits of what was known. He had some ideas on how to continue, but his initial computations showed that this path did not lead anywhere. Several weeks passed without finding a new approach. The mathematics needed to tackle the problem had not yet been developed. To go on, he would have to create it himself. He realized that this would take an enormous amount of time. Would he complete it by the prize deadline? It was difficult to estimate, but this seemed to be the only way—at least the only way he could imagine.

Months and more months of hard work followed. He abandoned his efforts many times, but always resumed them. After two years he no longer put the problem aside: he worked continually. It was a beautiful problem and Poincaré managed to construct a huge edifice. He produced much new mathematics solely to apply it to the three- and n-body problems. He invented the notion of *integral invariants* and used them to prove the *recurrence theorem*, which we shall describe later. He developed a new approach to periodic solutions and stability, including the idea of *characteris-*

tic exponents, which are now standard tools in dynamics. Some of his results were unexpected and ran counter to his intuition. He could not completely prove others, but felt sure of their validity. He had a mass of suggestive and partial results, far from the clarity he demanded of himself and his work. Unfortunately, a solution to the original problem still seemed far away, and each step forward gave him a clearer understanding of how distant the final goal must be.

Correspondence subsequently published in *Acta* shows that Mittag-Leffler encouraged his friend Henri Poincaré at several times during this period, even suggesting that he should submit answers to the other questions posed by the jury. But Poincaré concentrated all his talent on the first.

In the third year things began to crystallize. The three-body problem revealed one of its secrets. Poincaré had proved the nonexistence of *uniform first integrals* other than the known ones. This meant that the three-body problem could not be solved by certain quantitative methods. In terms of our earlier discussion, it does not reduce to a problem of lower dimension: even the restricted three-body problem, to be discussed later, would need two degrees of freedom for its full description. Poincaré's theorem came as an improvement of a similar result that had been published by the German mathematician Ernst Heinrich Bruns in 1887. Poincaré discovered and explained many other properties, and among them he believed he had found proof of a kind of stability result for the restricted three-body problem. He would not realize until later that this proof was wrong and that his attempts to correct it would lead him to a much more striking discovery.

He felt that these results were sufficient, for now he understood that the question was much more difficult than anyone could have expected. Mathematicians would have to change the way they thought and approached dynamical problems of this type. Quantitative methods, which sought explicit formulas and integrals, were too weak. He stood before a huge door, leading to a mysterious world. He opened it and paused on the threshold, looking in. For the first time in months he relaxed. "This problem will provide an inexhaustible source of results for future generations," he thought. "It is too much for me alone. All I might do is take a few more steps inside and that would be enough to occupy the rest of my life. At least I understand now that nobody will solve the problem alone. It cannot be comprehended by our generation."

Poincaré decided to write up his research. He no longer thought of the prize itself; he had to make his results known. Everyone, not only those working on the *n*-body problem, must understand the new point of view that he had developed. Quantitative, analytical methods were good only in part. *People should also begin thinking qualitatively and geometrically!* The fastest way to spread these ideas would be to submit his work to the contest, unfinished and imperfect though it was. That same evening he sketched the outline of his future manuscript. It was submitted on 17 May 1888.

Under the rules of the competition, entries had to be anonymous, being identified only by an epigraph. Recalling his childhood wonder at the night sky, Poincaré chose *Nunquam praescriptos transibunt sidera fines*: Nothing exceeds the limits of the stars.

On 21 January 1889 the prize was awarded to Jules Henri Poincaré, for his remarkable contributions to understanding the *n*-body problem and related fundamental questions of dynamics.

LES MÉTHODES NOUVELLES . . .

Within little more than a year, Poincaré's breakthrough became known worldwide. Several details had to be polished, but finally the paper was ready for publication. It was huge. The printed version, which finally appeared in volume 13 of *Acta Mathematica* for 1890, ran to 270 pages, the size of a modest textbook. And it was, in essence, the first draft of the great three-volume work to follow. At the end of this chapter we shall probe the story of this paper and its publication more closely; for the moment, it is enough to note that, after its appearance, Poincaré's name became a landmark in mathematics.

The importance and value of the new mathematical door opened in "Sur le problème des trois corps et les équations de la dynamique" is inestimable. His success gave Poincaré new energy and enthusiasm. He decided to write a book on this subject and started work immediately, his goal being to develop and extend the ideas of the prize paper. In 1892 the first volume of *Les méthodes nouvelles de la mécanique céleste* appeared. It was quickly followed by the second, published just one year later. The first volume treated *periodic solutions* and *the nonexistence of uniform integrals* as well as *asymptotic solutions* of the three-body problem. It also contained the first theorems on *continuous dependence of solutions with respect to a parameter*. The second volume focused on the perturbation series methods developed by Newcomb, Gyldén, Lindstedt, and Bohlin, and on their applications to the three-body problem.

Progress on the third and final volume of the book was not as straightforward as the first two. Poincaré's work on qualitative methods was delayed by other projects. Moreover, his reservations regarding some of his new discoveries made him cautious. The main topics he treated in this last volume were *integral invariants*, *stability in the sense of Poisson*, *periodic orbits of second kind*, and the so-called *doubly asymptotic solutions*, which he had introduced in the prize essay.

We shall subsequently explore several of these topics, but it is the *doubly asymptotic solutions* that now concern us. They appear in the final chapter of *Les méthodes nouvelles* and it was their behavior that obsessed Poincaré at the beginning of our story. For almost ten years he struggled to understand them. To appreciate the difficulties he encountered and eventually overcame, we must return to the theory of differential equations. The next three sections are rather technical and dry, but they are necessary if we are to provide a clear description of what a mathematician understands by "chaos." The reader who so prefers may skim these sections and pick up the story at "Pandora's Box," returning later to fill in details.

Fixed Points[*]

Remember that a solution of a differential equation can be represented by a curve in a plane, in a higher-dimensional space, or on a manifold. In a special case, this curve is a single point. If we recall the analogy with the surface of a river, the center of a vortex is such a point. Once caught, a scrap of floating wood is confined forever to the middle of the vortex, if the flow pattern does not change. This is not the only possible configuration: a *fixed point* can exhibit several distinct forms, some of which are illustrated in figure 1.6. Such an orbit is also called a *rest point* of the flow, or an *equilibrium solution* of the differential equation. The rates of change vanish at these points and the variables of the differential equation remain constant.

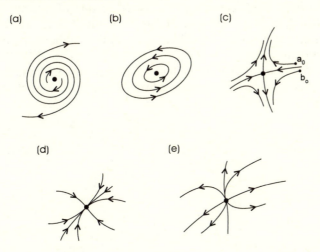

Figure 1.6. Possible behavior of the flow near a rest point: (a) spiral source, (b) stable point, (c) saddle point, (d) sink, (e) source.

As figure 1.6b shows, there may exist *closed* or *periodic* orbits around the fixed point. Some solutions may tend toward or recede from it in spirals, as in figure 1.6a, or directly, as in Figure 1.6c. It can also happen that all orbits in the neighborhood tend to the equilibrium (the so-called *sink* in fig. 1.6d) or leave from it, in case of a *source* as in figure 1.6e. The picture of figure 1.6c, in which two orbits approach, two leave, and all others skirt past, is called a *saddle point*.

Equilibria can be reached by nearby solutions only in infinite time. One either remains at an equilibrium state forever, or one tends asymptotically to it. Also, since speed slows to zero at an equilibrium, orbits skirting a saddle point spend a long time in its neighborhood, and in doing so get stretched far apart: imagine the fates of two points started close together but on opposite sides of the unique curve approaching the saddle (points a_0 and b_0 in fig. 1.6c). This latter curve is called a *separatrix*, since it separates orbits with different futures. (In pointing out that it takes a solution of a differential equation infinitely long to reach an equilibrum, we should stress again that we are describing a *mathematical model*: an idealization and simplification of physical reality. In the real world, friction typically stops a motion dead after only a finite time.)

What is the physical interpretation of an equilibrium solution? As we have pointed out, these states correspond to processes whose properties do not change with time. For example, a particle at rest (the fallen ball of fig. 1.1), or a particle moving uniformly on a line with constant velocity, may both be represented by equilibrium solutions. A pendulum at rest, in its lowest or highest position, provides two further examples of equilibria (see fig. 1.7). Clearly a pendulum consisting of a heavy bob fixed to a rigid rod, such as you will find in a grandfather clock, will remain forever in its lowest position as long as it is initially there and no exterior forces act on it.

In fact, the same is true for the highest configuration, although it is hard to imagine how one could in practice bring a pendulum to such an inverted

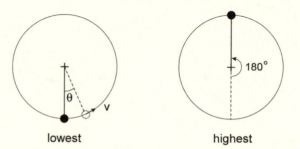

lowest   highest

Figure 1.7. The two equilibrium positions of the pendulum.

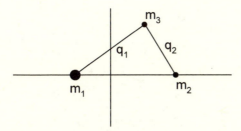

Figure 1.8. The planar restricted three-body problem.

position and keep it there without holding it. Any small deviation will cause it to fall to one side or the other. The notion of stability in the theory of differential equations addresses this issue. In the first case we say that the equilibrium solution is *stable*, whereas in the second case we call it *unstable*. We have stability when no solution leaves a neighborhood of the equilibrium, as in figures 1.6b and 1.6d. Stability, like a diamond, is forever: it refers to the long-term behavior of solutions. If a solution is stable, nearby solutions remain near it for all time. Instability occurs when at least one orbit leaves the equilibrium, even if it might eventually return to the same neighborhood; see figures 1.6a and 1.6c. We shall discuss stability at greater length in chapter 4.

Poincaré's problem concerned orbits that tend toward and recede from equilibrium solutions and periodic orbits. A satellite coasting in to park at the point where the earth's and the moon's gravitational pulls cancel each other would follow such an *asymptotic* orbit. Poincaré was studying the *planar restricted three-body problem*, a special case of the classical three-body problem in which all bodies move in the same plane and one mass is very small in comparison to the other two. The latter consequently follow the ellipses of the soluble two-body problem, as if the smaller mass were absent. Actually, he concentrated on the special case of circular orbits, for which one can pass to a corotating coordinate system in which the two larger bodies appear stationary. The position and velocity of the third mass may then each be specified by just two coordinates: a two-degrees-of-freedom system. Figure 1.8 shows the large bodies, m_1 and m_2, rotating about their common mass center, to which the coordinate system is attached. The position of the small mass, m_3, is given by the two interparticle distances q_1 and q_2.

Poincaré's computations showed that *doubly asymptotic* orbits could occur around certain equilibria for this system. These are curves that leave the equilibrium and then return asymptotically to it as time increases, as shown in figure 1.9a. In the final volume of *Les méthodes nouvelles*, he named such orbits *homoclinic*, since they *incline* to and from the same

(a)   (b)   (c)

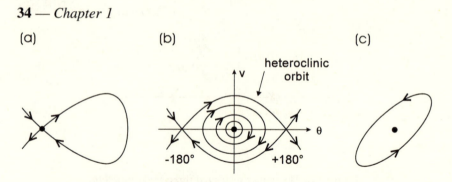

Figure 1.9. A homoclinic (a), a heteroclinic (b), and a periodic orbit (c).

limit. If a curve joined two distinct equilibria, he called it *heteroclinic* (fig.
1.9b). In this latter figure, we actually show the phase portrait for the pendulum of figure 1.7. The variables describing its state are θ, the angle the
rod makes with the vertical, and v, the angular velocity of the rod and bob.
The two equilibria referred to above appear as the stable fixed point (θ, v)
$= (0, 0)$, surrounded by periodic orbits correponding to small swinging
motions, and the unstable fixed point $(\theta, v) = (180°, 0)$ corresponding to
the highest or "upper" rest position. (The latter appears twice, at $(180°, 0)$
and $(-180°, 0)$, which both represent the same point in physical space, as
we shall see below.)

   In the mathematical idealization that leads to the differential equations
whose flow is shown in figure 1.9b, we assume a "perfect" pendulum, with
no loss of energy due to friction at the pivot or air resistance. Once set
swinging with a certain amplitude, it maintains that swing forever. Gravity
alone acts as the restoring force: released from any position between the
equilibria, the pendulum falls down, gathering speed, and then swings back
up to the same level on the opposite side, where it comes to rest again for
an instant before repeating its swing in reverse. Hence all these orbits are
closed. The heteroclinic orbit correponds to a motion in which the bob,
infinitesimally disturbed from its unstable upper position, swings down,
picking up kinetic energy, which it then surrenders as it exactly returns to
its starting point, taking infinitely long to do so.

First Returns[*]

   Recall that a periodic orbit such as the one in fig-
ure 1.9c, in spite of having a phase portrait similar to the homoclinic one,
has a different interpretation. First, there are no equilibrium points on it,
and second, the state takes only a finite time to complete its circuit, which
it then repeats ad infinitum. On a homoclinic curve, the state moves from

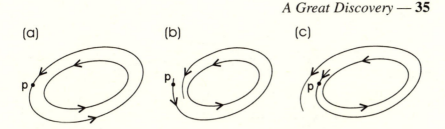

Figure 1.10. Three possible behaviors of a solution curve outside a periodic orbit.

the rest point in the distant past and does not return until infinitely far in the future, having passed along the curve only once.

Poincaré's analysis started around a periodic orbit. Due to continuity of solutions with respect to initial data, curves starting at points close to the periodic orbit must stay close to it for a certain interval of time. Let us take a point p, not on the periodic curve, but close enough to it so that the solution curve passing through p does not escape before one circuit is completed (see fig. 1.10). How does this orbit behave? We have three possibilities: (1) the curve through p may be closed as in figure 1.10a, so that it forever keeps its distance from the periodic orbit; it may spiral (2) toward or (3) away from the given periodic trajectory, as in figures 1.10b and 1.10c. In the case of approach, the orbit through p takes infinitely many circuits to actually reach the periodic orbit, and similarly infinitely many circuits are required "in the past" for the receding orbit.

These are indeed the only possibilities. Different behavior might occur if either the curve through p intersected itself or other solution curves, or if it left a neighborhood of the closed curve too fast. But the first possibility violates the uniqueness property, and the second contradicts continuity of solutions with respect to initial data. If a solution starts sufficiently close to the closed orbit, it must make at least a few circuits in the same neighborhood.

Poincaré imagined a simpler way to look at such orbits. He drew a *cross section D* through any point on the periodic orbit, so that D contains no equilibria; we show this in figure 1.11. Here D is a *transversal* line to the periodic orbit and to nearby curves. Transversal means that it cuts cleanly across the solution curves and does not glance them tangentially. Take the point q on the cross section, where the closed trajectory crosses it, and follow the orbit in the direction given by the arrow of increasing time. Naturally, we next hit the cross section D again at the same point q. Continuing our trip along the periodic orbit, we meet the section D once more. The encounter takes place again at q. Repeating the process, we always strike the same point q of the cross section D.

What happens if we start our walk from a point outside the closed curve, say o? We first meet the cross section at o_1, the second time at o_2, the third

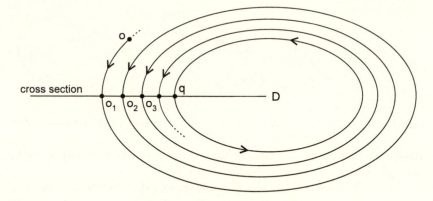

Figure 1.11. The first return map.

time at o_3, and so on, approaching (or receding from) the periodic orbit at every circuit. Poincaré realized that it is easier to study such sequences of points in place of the whole solution curve. The analogy becomes clear. If the ordering of the points o_1, o_2, ... o_n, ... on the section D is toward the periodic orbit, as in figure 1.11, then the behavior of the spiraling orbits is the one in figure 1.10b, and conversely.

The behavior of orbits returning to the section is described by what we now call the *Poincaré* or *first return map*. This is a function assigning to each point in D its unique *image* or *iterate*: the point at which the solution next returns. If we denote the function by F, the first iterate of F at o_1 is written as $F(o_1) = o_2$, the second iterate is $F(F(o_1)) = F(o_2) = o_3$, and so on. The map can be seen as an input-output relation: given the present position o_n, it predicts the next encounter $F(o_n) = o_{n+1}$ with the cross section. Each iterate implies another trip around. The infinite collection $(o_1, o_2, o_3, \ldots)$ of all iterates of F at o_1 is called the *orbit* of the Poincaré map through o_1.

Let us take a closer look at this important mapping. First note that the two-dimensional problem describing the structure of curves in a plane is reduced to a one-dimensional problem, the study of points mapped on a line. The point q on the periodic orbit becomes a fixed point for the first return map. In fact, the setting is only apparently *discrete* (i.e., in terms of points instead of continuous lines). Taking a point r close to o, one gets another sequence of image points, $r_1, r_2, \ldots, r_n, \ldots$, which tend to q. The iterates of all points on a segment connecting r and o join to make the set of points on the cross section D become a continuous segment. Thus, the geometrical study of the first return map is superficially quite similar to that of the curves representing solutions of the underlying differential equation. Any set of points forming a smooth curve that is mapped into itself by F is called an *invariant curve*. The solution curves shown earlier in this chapter are examples of invariant curves. If one takes any collection of initial conditions on such a curve and follows the solutions through those points to their images by iterating F, one remains on the curve.

There is, however, a major difference between curves describing solutions of a differential equation and invariant curves obtained from first return maps. As we have seen, solution curves of differential equations cannot intersect each other due to the uniqueness property (if they did so, the crossing point would have two possible futures and two pasts, like the curves $r_{-1} \, r_0 \, r_1$ and $s_{-1} \, s_0 \, s_1$ in fig. 1.2). Orbits of maps, however, are sequences of points, like those marching along the cross section in figure 1.11, so this observation does not apply to invariant curves of first return maps. They may indeed intersect each other transversely.

Poincaré maps can also be defined in higher-dimensional phase spaces. For curves defining solutions of a differential equation in a three-dimensional space, the cross section is two-dimensional and the corresponding first return map also has two dimensions. The Poincaré map always reduces the dimension of the phase space by one.

A Glimpse of Chaos[*]

The next few paragraphs may be heavy going, but they will allow us to get a little way into Henri Poincaré's mind and show how manifolds, introduced above, are used to help understand the structure of phase space. (Thus we will be better able to appreciate how natural it was for him to contribute important ideas in topology as well as to applied dynamics.) However, we recognize that the reader coming to these ideas for the first time may prefer to skip the details; after all, it took Poincaré himself many years to come to terms with them! For those readers, we note that the picture of *chaos* in Poincaré's imagination was similar to that of figure 1.14, below. The interlaced curves in that figure represent solutions of the underlying differential equations which approach a particular periodic solution (p) in the future (S), and in the past ($\mathcal{U}$). Their intersection points correspond to solutions which are trapped in both the distant past and the future, but which can perform a "chaotic" dance in between.

To appreciate the details of Poincaré's insight, let us return to the equation describing the motion of the pendulum of figure 1.7, but with the addition of a new effect. Suppose that the pivot is periodically shaken, imparting a chain of impulses, much like a child moving its legs can pump up the oscillations of a swing. Now the law governing the motion varies with time. It is still Newton's second law, but from the pendulum's point of view, the gravitational restoring force is time dependent. So now we need three quantities to fully describe the state of the system: its angular posi-

(a)   (b)

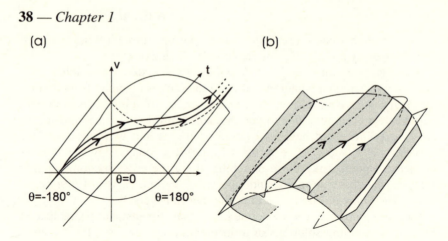

Figure 1.12. The three-dimensional phase space of the pendulum, (a) without and (b) with periodic shaking.

tion, θ, its velocity, *v*, and *time* itself, *t*. These three coordinates describe an *extended* phase space. If the periodic motion is small, the "up" and "down" equilibria will survive as small periodic motions themselves, so that, in the three-dimensional phase space, they look something like figure 1.12b. Note that, in this picture as in figure 1.9b, the unstable motion appears twice: near θ = 180° and θ = −180°. Both curves refer to the same motion in physical space since the angles 180° and −180° specify the same thing: you end up facing backwards whether you make a half turn to the left or to the right.

Now, Poincaré was actually studying the restricted three-body problem of figure 1.8, but his two-degrees-of-freedom system had a conserved quantity. As we saw earlier, its solutions were therefore restricted to a three-dimensional manifold. In it, they appear much like those of the simpler pendulum equation. Indeed, in the final version of his prize-winning paper, Poincaré himself used the pendulum as an illustrative example.

Returning to the pendulum, let us focus on the upper (unstable) motion and suppose for a moment that the periodic force is turned off, so that it reverts to being an equilibrium position. This appears in the extended phase space as a straight line (fig. 1.12a). If the bob is displaced, say to the left, and given *precisely* the right initial velocity, it will swing around once and approach the unstable equilibrium again, taking infinite time to do so. The full set of such motions forms a two-dimensional *stable manifold*, and there is similarly a two-dimensional *unstable manifold* filled with orbits that originated at the unstable equilibrium infinitely long ago (to mathematicians, time can equally well run backwards as forwards, and so we can say "orbits which approach the equilibrium as time goes to minus infinity").

For the unperturbed pendulum, energy is conserved. The system therefore enjoys a symmetry: its behaviors in the past and future are identical. This implies that the stable and unstable manifolds join smoothly to form a sheet looping from the unstable equilibrium and back to it, as shown in figure 1.12a. In fact, there are two such sheets: one corresponding to orbits in which the pendulum makes a clockwise turn, and the other, those in which it makes counterclockwise turns.

These stable and unstable manifolds survive, albeit in deformed fashion, when the periodic force is turned on again. In figure 1.12b we try to suggest how this happens. Energy is no longer constant, for we are pumping the pendulum, and so the future fate or past history of an orbit depends crucially on the timing with which it approaches the unstable motion. Arbitrarily small changes can cause the bob to pass over the top or to swing back, and this is reflected by the fact that the two manifolds now wrinkle up and split apart. They continue to intersect only on those distinct orbits for which the timing is such that, on average, energy input exactly balances output so that the solution may end where it began. (To prove that any such "balanced" orbits persist, one must do a calculation in which one averages the energy equation along orbits.) The picture rapidly gets very complicated, and it is here that the idea of a cross section and a return map can *really* help.

But before we can take a cross section, we must represent the phase space in a different manner than that of figure 1.12, in which we have imagined time as a line, running on forever, as may seem natural to (human) minds. In the limited world of the pendulum, however, time is only explicitly present in the motion of the pivot, and so time might as well be periodic. If the motion repeats on a cycle of duration T seconds, then the state specified by the three quantities θ, v, and t is indistinguishable from that specified by θ, v, and $t + T$ or θ, v, and $t - T$. This implies that the orbits, manifolds, and all else in figure 1.12b are also periodic with period T, and so in our phase space we need only include one "frame" of duration T in the time direction, as indicated in figure 1.13a. In this periodic phase space, we start an orbit with initial condition (θ_0, v_0) at time $t = 0$ and follow it for one cycle to (θ_1, v_1) at time $t = T$. The behavior during the next cycle is then obtained by restarting at (θ_1, v_1), with t reset to 0, and following the orbit for another cycle, to (θ_2, v_2). This can be repeated ad infinitum.

Topologically, we *identify* the cross sections at $t = 0$ and $t = T$, regarding them as a single plane, and so the orbits or flow of the differential equation induce a Poincaré return map P on this plane, just as in the example of figure 1.11, in which a mapping was induced on the cross section line D. As in that example, an orbit of the differential equation (a curve) becomes a sequence of distinct points for the map. See figure 1.13b. P takes (θ_0, v_0) to (θ_1, v_1), (θ_1, v_1) to (θ_2, v_2), etc., and now the two-dimensional stable and

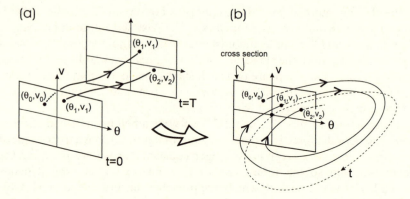

Figure 1.13. (a) The periodic phase space, and (b) the cross section and the Poincaré map.

unstable manifolds of figure 1.12 become invariant curves for the Poincaré map (imagine slicing the sheets of fig. 1.12 with the planes of fig. 1.13). We will denote these two curves by S and U, respectively. *Invariant* means that the image under the map, $P(C)$, of any piece C of S or U, lies in S or U, respectively.

Since it repeats with exactly period T, the (unstable) periodic motion appears as a fixed point of the map, which we shall call p. The curves S and U belong to the (two-dimensional) stable and unstable manifolds of this periodic motion, and so points on them march along toward p under the action of P and its inverse P^{-1}, respectively. As we have stated, due to the periodically varying force we generally expect the stable and unstable manifolds to wrinkle and split, so that S and U will appear as two separate curves on the cross section, unlike the single smooth curve that character-izes the energy-conserving, unperturbed pendulum. However, as we have seen, for some special orbits we still expect these curves to intersect. As in the unperturbed case, they separate orbits that pass "inside" and "outside" the unstable inverted equilibrium, corresponding to motions in which the pendulum makes (large) swings back and forth and those in which it makes full overturning rotations.

We are finally in a position to describe the phenomena Poincaré was struggling to understand when we followed him out on his walk. His calcu-lations had led him to the conclusion that, in the three-body problem, the analogues of the curves S and U intersected transversely at a point q_0. Now, as we have said, the point p is fixed, the map P leaves it unchanged, but starting at q_0, P generates an orbit of distinct points $P(q_0) = q_1$, $P(q_1) = q_2$, . . . in forward time, and in backward time $P^{-1}(q_0) = q_{-1}$, $P^{-1}(q_{-1}) = q_{-2}$, ... (Each of these points—the images of q_0 under P—is specified by two coordinates; strictly, we should write $q_n = (\theta_n, v_n)$.) Since all points on S

march toward p as time increases, the sequence q_n approaches p as n tends to infinity, and similarly the sequence q_{-n} approaches p as n tends to infinity. The homoclinic intersection point q_0 belongs to both S and U, so we conclude that *one transversal homoclinic point produces infinitely many more transversal homoclinic points*. These are Poincaré's "doubly asymptotic" points.

There is more to come. Along with the images and preimages of q_0, the map carries images and preimages of the region near it and, in particular, of arcs contained in the curves S and U. Now these curves are *oriented*: there is a distinguished direction on them—nearer to or farther from their origins at the fixed point p—indicated by the arrows in figure 1.14a. The Poincaré map preserves orientation, so that if S crosses U from left to right at q_0, their images will do the same at every point q_n. Connecting up the little arcs near these points to fill out the curves, we see that there must be at least one additional intersection r_n between each point q_n and its image q_{n+1}. These two infinite sequences q_n and r_n are called *primary homoclinic orbits*, and we sketch the situation in figure 1.14(a). They correspond to motions of the perturbed pendulum in which the bob makes a single circuit before settling back to the unstable inverted state.

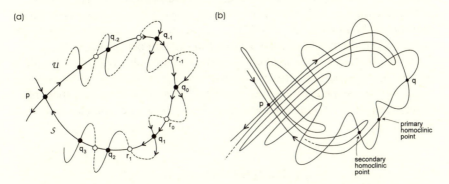

Figure 1.14. A first (a), and (b) a second approximation to the homoclinic tangle.

There is yet more. As indicated in our sketch, the little arcs of U and S, between the primary homoclinic points, get mapped closer and closer to p under iteration of P forwards and backwards, respectively. Approaching this saddle point, they come under the influence of its unstable, stretching behavior. These little loops are therefore stretched out along the directions of the curves U (under forward iteration) and S (under inverse iteration), as indicated in figure 1.14b. This leads to further intersections and *secondary homoclinic orbits*, corresponding to the pendulum swinging twice around before settling down.

Now we are in the mathematician's playground and we can repeat this argument ad infinitum to get tertiary and quaternary homoclinic orbits, and indeed orbits of any temporal (or geological) type! So we find an infinite collection of homoclinic orbits, each of which contains an infinite sequence of homoclinic points. Starting with a single intersection point of the stable and unstable manifolds S and U, we have found infinitely many infinite families of such points, pinning the curves together in a *homoclinic tangle*, a sort of web or trellis which Poincaré, showing more insight and caution than ourselves, did not attempt to draw. Each thread of the web is an arc belonging to S or to U, and therefore these threads separate orbits that eventually pass to the left or to the right of the saddle point p in the future, or that originated on different sides of it in the past. With infinitely many separatrices packed into a finite region of phase space, it is practically impossible to predict the fate of an orbit starting at any given initial condition.

In the final volume of *Les méthodes nouvelles*, at the beginning of section 397, Poincaré describes the homoclinic tangle: "When we try to represent the figure formed by [the stable and unstable manifolds] and their infinitely many intersections, each corresponding to a doubly asymptotic solution, these intersections form a type of trellis, tissue or grid with infinitely fine mesh. Neither of the two curves must ever cut across itself again, but it must bend back upon itself in a very complex manner in order to cut across all of the meshes in the grid an infinite number of times."

As we shall see in the next chapter, the infinite collection of homoclinic points and tangled separatrices is only a small part of the story. Much more remained to be uncovered, but Poincaré's *homoclinic tangle* appears to have been the first mathematical manifestation of the phenomenon now called *chaos*.

PANDORA'S BOX

Poincaré was so struck by his discovery that it was some time before he could accept the idea that a homoclinic tangle might exist. Ten years before, he had been the first to appreciate the difficulty of the *n*-body problem. As we will soon see, he had already pointed out the possible existence of transversal homoclinic orbits in his prize essay, but at that time he went no farther in trying to understand the implications of these orbits. The discovery he eventually described in the third volume of *Les méthodes nouvelles* suggested that there would be little chance of ever completely solving the problem. It seemed hopeless for him and his contemporaries, or even for future generations, to seek a general solution. Besides, it was hard to accept the idea of chaos. Such a concept went counter to the philosophy of his epoch.

Differential equations, including those modeling physical phenomena, describe deterministic processes. The existence and uniqueness properties guarantee that, once the initial condition is known, the solution is completely determined. The present predicts the future, at least as far as the mathematical model is concerned. One's only problem is to find (or approximate) this solution. Poincaré's work had changed nothing in this respect, but he had shown that, in many cases, the chances of finding explicit solutions were practically zero, and that, if found, they would depend with exquisite sensitivity on the initial data. How could he explain to his contemporaries that their quantitative methods would often encounter insurmountable barriers, that determinism does not imply accurate prediction? Who would accept this? How could he convince the scientific world that, over sufficiently long intervals of time, the gravitational motion of just three celestial bodies might be as unpredictable as the weather?

The intellectual spirit at the end of the nineteenth century was aggressive and optimistic. The ground had been prepared for a revolution in science and technology. The great international expositions, such as that for which Alexandre-Gustave Eiffel's tower had been built, had convinced the general public that human intelligence and ingenuity could solve all social and economic problems. A new age was dawning. Poincaré's reservations could not deflect the cornucopia of scientific discoveries just over the horizon. His work ran counter to the mood of the times, and it seemed unlikely that many would accept it. It is probable that Poincaré himself failed to completely understand the deep consequences of his conclusions and all their subtleties. He was a mature, well-established researcher, reluctant to open this Pandora's box. He had found it and that was enough. His philosophy would permit him to go no further.

In fact it was not until the 1980s, when personal computers and workstations with graphic displays became widely available, that the general scientific public would accept chaos as a pervasive phenomenon.

We are fairly sure that things happened in this way. A well-documented event a few years later lends credence to our supposition: the birth of *special relativity*. Poincaré had the data and premises to create that theory himself. It is known he also had the idea of the relativistic axioms; but they appeared so unusual and counter to scientific common sense, that he rejected them from the outset. Albert Einstein, a clerk in the Swiss Patent Office, young and completely unknown in higher scientific circles, had nothing to lose. He dared and succeeded. There were, however, many examples of researchers who had tried and failed. Their names and theories have faded from memory, as Poincaré well knew.

The final volume of *Les méthodes nouvelles de la mécanique céleste* appeared in 1899. Poincaré did not sketch the homoclinic tangle described above. He merely noted at the end of the book that such a picture was too complex to be drawn, and added that these facts can give us a good idea of how difficult most problems in dynamics are.

POINCARÉ'S MISTAKE

The story so far seems fairly straightforward; it is the version that most scientists accept. But recent evidence shows that things were not quite so simple. An unusual chain of events lies behind the neat accounts offered by the conventional histories. We will now describe what is known today.

In a letter addressed to Mittag-Leffler, dated 16 July 1887, when he was hard at work on the prize question, Poincaré claimed to have proved a stability result for the restricted three-body problem. "In this particular case," he writes, "I have found a rigorous proof of stability and a method of placing precise limits on the elements of the third body . . . I now hope that I will be able to attack the general case and that, from now until the first of June, I will, if not completely resolve the question (of this I have little hope), then at least find sufficiently complete results to send in to the competition."

It is not at all clear which stability result Poincaré is speaking of. The letter was published in *Acta Mathematica* not until after his death. A footnote by the editor (most probably still Mittag-Leffler) suggests that Poincaré is referring to his *recurrence theorem*, which appears in the published paper and which we shall discuss in chapter 4. However, this seems unlikely, since Mittag-Leffler already knew that the recurrence theorem works in much more general settings. We may therefore suspect that the claim of stability describes a quite different situation, namely the one supposed to be Dirichlet's discovery. Such a stability statement would have been in line with the prize question as it was proposed.

As a referee of Poincaré's paper, in his report to Mittag-Leffler, Weierstrass asserts: "I have no difficulty in declaring that the memoir in question deserves the prize. You may tell your Sovereign that this work cannot, in truth, be considered as supplying a complete solution to the question we originally proposed, but that it is nevertheless of such importance that *its publication will open a new era in the history of celestial mechanics* [our emphasis]. His Majesty's goal in opening the contest can therefore be considered attained."

On the other hand, it seems that Weierstrass was in poor health when he refereed the paper. In fact, his report on it was not actually submitted until after the king had approved the award. In a letter addressed to his friend and former student Sonja Kovalevskaia (whom we shall meet again in chapter 5), the German mathematician wrote: "As a judge I have been unable to correct some possible errors in Poincaré's work, but I have submitted annotations to the King, asking that *Acta Mathematica* print those annotations in the same issue as Poincaré's work."

These reservations never appeared in the journal. In his role as editor, Mittag-Leffler instead added a note explaining that Weierstrass's poor health made publication of his evaluation impossible. In addition to Poincaré's memoir, volume 13 of *Acta Mathematica* contains only a paper by another French mathematician, Paul Appell, who won the second prize for his work on one of the other problems proposed by the jury. Appell's paper was published together with an evaluation of it by Charles Hermite, the third judge.

Several questions arise. Why was Weierstrass's report not published? What happened to the stability result Poincaré had claimed to have proved in 1887? There is no such statement in the printed version of the published paper. What lies behind this story?

We now know that, in May 1888, Poincaré submitted for the contest a paper in which he claimed to have shown certain stability results for the restricted three-body problem. The paper was duly refereed by Mittag-Leffler and Weierstrass, the former traveling to Germany so that they could read all the entries together. It was awarded the prize in January 1889 and preparations began for publication in *Acta Mathematica*. The long process of typesetting and printing ran from April to November 1889. During this period, Edvard Phragmén, an editor of *Acta* whom Mittag-Leffler had appointed as a preliminary reader of the entries and who was then copyediting Poincaré's memoir for publication, raised questions about certain unclear passages (some of these concerned small denominators, which we shall discuss in chapter 5). Mittag-Leffler communicated the questions to Poincaré, who began responding, writing extensive explanatory notes to add as appendices to the main text. Then, at some time in the fall, Poincaré realized that he had made a more serious mistake in another part of the paper. On 30 November he telegraphed Mittag-Leffler to stop printing immediately, pending the arrival of a letter explaining the error.

In *Newton's Clock* (1994), Ivars Peterson states that Phragmén *found* Poincaré's error. In this he follows others, including Moulton, whom we quote below. However, it now seems more likely that, in drawing attention to other points, Phragmén primarily did "permit [Poincaré] to discover and correct a serious mistake [*importante erreur*]," as claimed in the introduction to the revised paper. Writing to Mittag-Leffler on 1 December 1889, Poincaré speaks of having "[told Phragmén] of an error I have made," rather than the opposite. He remained reticent about the error itself; his direct responses to Phragmén's questions all concern other, subsidiary matters. Nonetheless, following the discovery and study of Poincaré's annotated copy of the original printed version, we can now describe the nature of the mistake.

Recall the discussion of how the stable and unstable manifolds of the saddle point of the pendulum split when it is subjected to periodic perturbations, as sketched in figures 1.12b and 1.14. Poincaré realized that perturbations would shift the analogous unperturbed invariant curves in the three-body problem and had developed series expressions to approximate their locations, expanding in terms of a small parameter. He correctly argued that the perturbed curves must continue to intersect, for otherwise, if one lay entirely "inside" the other, areas would not be conserved by the iterated mapping. This would in turn violate the law of conservation of energy. He considered various possibilitites, and in subtle perturbation calculations he eliminated them all. He concluded that the perturbed curves must therefore coincide exactly. If this were the case, it would imply that the corresponding two-dimensional invariant surface for the flow prevented solutions from escaping, as does the separatrix for the unperturbed pendulum in figure 1.12a. But Poincaré failed to recognize the possibility of transversal intersection of these curves and the attendant phenomenon of chaotic motions. He had therefore drawn the incorrect conclusion of stability.

This was not the only problem that faced Mittag-Leffler. Shortly after the prize award had been announced, but before the mistake was discovered, another colleague and fellow editor of *Acta*, the astronomer Hugo Gyldén, announced that, in a paper published two years previously, he had anticipated Poincaré's (premature) claim of stability. On being informed of this by Mittag-Leffler, Poincaré had replied that he found Gyldén's paper difficult to read and inconclusive, but he believed that convergence of certain series had not been proven in it. It seems that Gyldén's "physical" approach was not sufficiently rigorous to satisfy either Poincaré or Hermite, to whom Gyldén also appealed. (In another paper published in *Acta* several years after the latter's death, and after correction of his own work, Poincaré was to argue that Gyldén's approach had indeed been wrong.)

Gyldén's challenge led to some troublesome debates in the Swedish Academy of Sciences, which Mittag-Leffler was not entirely able to quell. In fact, the controversy continued to rumble for some time. As late as 1904, following a debate on the history of celestial mechanics, Hugo Buchholz, a former student of Gyldén, published a paper defending his adviser's work. In it he acknowledges that Poincaré did prove divergence of the asymptotic series in certain cases, but argues that this does not invalidate

Gyldén's claim in other cases. He also admits that Gyldén's paper is very hard to read: a fact corroborated by Richard McGehee (see the next section), who more recently attempted to understand it, without much success. In any case, as the spring of 1889 advanced and the furor seemed to be abating, the "real" mistake was found.

We can imagine Poincaré's chagrin when he realized the nature of his mistake. In the letter we quote from above, he goes on to say: "I will not conceal from you the distress this has caused me." To have created an entirely new way of looking at dynamics and then overlooked such an obvious point! Yet research is often like this: the subtle and the obvious are intermingled, little is familiar, and such errors are more common than one might imagine. This is why new mathematical and scientific ideas and the papers that describe them are usually subject to intense review and criticism by other scientists before publication. When mistakes are found, the researcher must swallow his pride and attempt to correct them. It is a common, almost typical procedure. But in this case the prize had already been awarded and so the stakes were unusually high.

Under enormous moral and temporal pressure, Poincaré did correct his mistake over the next few months. This entirely changed the nature of his paper and it would change the whole course of celestial mechanics and dynamical systems theory. He realized that the intersecting invariant curves did not prevent orbits of the mapping from escaping, that the system indeed might be highly unstable, and that proof of this would actually *follow* from the transversal intersection of curves corresponding to a homoclinic orbit, as we described earlier. In adding explanations of other points that Phragmén had raised, correcting his mistake, and beginning to draw the startling conclusions from it, the 158 pages of his original paper grew to 270. In January 1890, a full year after award of the prize, he submitted the revised manuscript, which was to appear in *Acta* as the paper we now know. In the introduction, evidently at Mittag-Leffler's request, he acknowledges his debt to Phragmén, but without explaining the nature of the error. Meanwhile, he refunded the 3,585 crowns and 63 öre, which the original printing had cost, thereby spending considerably more than his prize money.

Poincaré's massively revised paper was printed between April and October 1890 and published late in that year, although the title page still bears the date of the prize award: January 21, 1889. The paper represented a major step toward the first example of chaos. The results obtained in the few months following his discovery of the mistake are indeed remarkable; they show Poincaré's powers of ingenuity and comprehension. The verdict of mathematical history is clearly on the side of the revised paper, but the fact remains that the prize was actually awarded to the earlier one, with its mistake. Of course, many ideas and results of great value survived unchanged from that version. In fact, as Forest Ray Moulton, a contemporary of Poincaré, a professor at the University of Chicago and a well-known researcher in celestial mechanics, wrote in 1912:

> It has been remarked a number of times, recently, that the original memoir of Poincaré on the problem of three bodies, for which the prize of King Oscar II was awarded, contained an error, and that the published paper differed from the one originally submitted. Unfortunately and erroneously the impression has been left in some of these statements that the first investigation was wrong and of little value. The original memoir did contain an error, which was discovered by Phragmén, of Stockholm, but it affected only the discussion of the existence of the asymptotic solutions; and in correcting this part Poincaré made no attempt to conceal the facts, and confessed fully his obligations to Phragmén.
>
> While the error was unfortunate, there is not the slightest doubt that in spite of it, and even had it been generally known at the time, the prize was correctly bestowed. If all the parts affected by the error are omitted, the memoir still remains one whose equal in originality, in results secured, and in extent of valuable field opened, it is difficult to find elsewhere. There are but few men, even of high reputation, who have produced more in their whole lives that was really new and valuable than that which was correct in the original investigation submitted by Poincaré.

A Surprising Discovery

Djursholm is a small, quiet town a few miles north of Stockholm. It is a pleasant place to visit, close to the strait and yet sufficiently far from the bustle of the city. Mornings are often foggy, and one can breathe the salty air of the sea. Some streets have changed little in the last hundred years. A sense of history permeates the narrow lanes.

A huge villa, almost a palace, houses the research institute of mathematics. This beautiful building is the former residence of Gösta Mittag-Leffler, whose name the institute bears today. In 1985 Richard McGehee from the University of Minnesota in Minneapolis spent part of his sabbatical leave at this august institution. He became interested in the work and life of Hugo von Zeipel (whom we shall meet in chapter 3), and found that documents preserved by Mittag-Leffler were useful to him. Mittag-Leffler's correspondence was alphabetically ordered and filed, along with carbon copies of the responses. The Swedish mathematician certainly knew how to preserve his papers for future generations.

But some documents in the archives had not been cataloged or arranged. One day McGehee opened a dusty box apparently containing several copies of volume 13 of *Acta*—that in which Poincaré's prize essay had been printed. He had a sneaking suspicion that this might be interesting, so he took a copy and inspected it in detail. He was amazed. To make sure, he took the same volume from the library's regular collection. Comparison of the two confirmed his suspicion. The two printed texts of Poincaré's paper were different.

What had remained generally unknown up to McGehee's accidental discovery was that the original manuscript, claiming stability, had indeed been published as volume 13 of *Acta Mathematica*. A few people, including Jürgen Moser (whom we shall meet in chapter 5), and the Swedish mathematician Lennart Carleson, did know of this, and it was actually a remark of Carleson's that had prompted McGehee to explore the archives. Printing was completed in mid-November 1889 and preliminary copies had been circulated to prominent mathematicians and astronomers. But after Poincaré's error came to light, Mittag-Leffler stopped distribution, recalled all the printed copies, and subsequently replaced them by the ones that we find in libraries today, containing the revised and corrected paper. McGehee had chanced upon some of the recalled volumes, which had probably escaped destruction due to secretarial negligence. (One of the copies is marked, in Swedish: "The whole edition was destroyed. M.L.") Following the discovery, scholars have examined the archives in more detail, and a copy of the original printing, amended by Poincaré with his answers to Phragmén's questions, has enabled us to reconstruct the events described in the previous section.

The decision to replace an issue after its publication is unusual for the editor in chief of a scientific journal. We do not expect scientific history to suffer the same revisions as political history under a totalitarian government. But perhaps Mittag-Leffler's reasons are understandable. First, he had to defend the reputation of the prize, of the journal, and of the discipline itself in the eyes of nonmathematicians. This is not an easy task, knowing how ineffective most scientists are as politicians. He had also to defend his own position as a leader of Swedish mathematics, as a person with influence in high society, and as an adviser of the king.

It is difficult and probably unfair to pass judgment on events a century ago, about which we have incomplete information. One can raise questions of ethics in science, but we should remember that Poincaré fully deserved the prize and that Mittag-Leffler's move was not intended to hide the mistake within the scientific world. This strange history adds a very human touch to the story, and we can still appreciate the positive influence of King Oscar's prize and Poincaré's contribution, without which the mathematics of our present century would have been very different.

Gösta Mittag-Leffler played a crucial role in the mathematical world by providing a bridge connecting the nineteenth and twentieth centuries. French, German, and English scientists had dominated the previous three hundred years, while Swedish mathematics was practically nonexistent. With his connections, Mittag-Leffler succeeded in creating a mathematical school, in raising funds to attract famous figures to his country, and in laying the foundations of one of the finest mathematical journals. And indeed it is remarkable to see how such a small country has been able to produce so many exceptional mathematicians in the short time since Mittag-Leffler. He must have been convinced that his actions were both just and necessary to defend his dreams and ambitions at a moment of political danger to all his considerable achievements.

Our story of Henri Poincaré's discovery of chaos in Isaac Newton's model of the solar system is almost over. Although he continued to return to problems in celestial mechanics and dynamics after finishing his great treatise on celestial mechanics, Poincaré moved on to other interests, including the authorship of delightful, lucid books on the nature of mathematical creativity and scientific reasoning. His election to the French Academy in 1908 was in fact based on the literary quality of these popular essays.

For the next thirty years, very few scientists were able or chose to build on his insights in dynamics. Prominent mathematicians such as Hadamard and Cartan developed some of Poincaré's ideas, but largely ignored chaos. It ran counter to the prevailing optimism, with its desire to control the natural world. In physics, the revolutions of relativity theory and quantum mechanics quickly replaced any interest in classical Newtonian mechanics among the research community. In mathematics, David Hilbert's program of rigorous axiomatization and development, quite counter to Poincaré's intuitive approach, began its long domination of the field. Yet Henri Poincaré had sown the seeds of a revolution as important as those that created modern physics. In the following chapters, we shall meet some of the people who picked up and continued his work through the twentieth century. We will also go back in time to recall earlier developments and to fill in other parts of the story.

2.

Symbolic Dynamics

The importance which Birkhoff attached to symbolism
in dynamics is made evident by the fact that his last
papers were largely concerned with the search for a
general symbolism characterizing a dynamical system.
—Marston Morse

WHEN HE ARRIVED at the department, George
Birkhoff found the letter in his mailbox. He had been expecting it for some
time and was eager to read it. He left the secretary's room, trying not to
hurry, and in the privacy of his office opened the envelope. Hands trembling, he unfolded the paper and read the first sentence.

With difficulty he suppressed a cry of joy. He had been fortunate. Harvard was offering him a professorship. Now the future would be financially secure and he would have sufficient time and peace of mind to continue his research. He had hoped for this for some years, but now that his dream had come true, he could scarcely believe it.

George David Birkhoff was twenty-seven years old when he received that letter in Princeton. Even though at the time he held only a temporary post there, the university did not want to lose this promising young man. Princeton's dean of the faculty offered Birkhoff an assistant professorship as soon as he learned of Harvard's interest. Two years before, in 1909, Birkhoff had arrived as a preceptor, a position similar to an instructorship at other universities. Now, with this promotion, he received his deserved recognition. For the moment, he decided to remain at Princeton. He would reconsider Harvard's offer the following year, and ultimately Birkhoff did move to Cambridge, Massachusetts, in 1912.

A Fixed Point Begins a Career

Birkhoff was born to Dutch parents in Overisel, Michigan, in 1884. He studied at the Lewis Institute between 1896 and 1902, and for another year at the University of Chicago. He received his Bachelor's degree in 1905 from Harvard. Two years after returning to

Plate 2.1. George David Birkhoff. (Courtesy of the American Mathematical Society)

Chicago he was awarded a Ph.D. in mathematics with the distinction summa cum laude. His doctoral dissertation dealt with the asymptotic character of solutions for certain linear differential equations depending on a parameter. This study marked the beginning, not only of Birkhoff's own remarkable career, but of the whole school of dynamical systems in the United States.

Birkhoff had great respect for Poincaré's achievements. He had carefully studied the French mathematician's works on dynamics. Birkhoff's first important result, which was to make him world famous, had its origin in Poincaré's last paper, "Sur un théorème de géométrie," published in 1912 in the Italian journal *Rendiconti del Circolo Matematico di Palermo*.

After some illness and surgery in 1908, Poincaré had seemed to make a complete recovery. In 1911, however, he had a presentiment that the end of his life was near. On 9 December he submitted to the *Rendiconti* an unfinished manuscript. In an accompanying letter to the editor, he expressed his worries about the problem on which he had worked for the past two years, saying that "at my age, I may not be able to solve it, and the results obtained, which may put researchers on a new and unexpected path, seem to me too full of promise, in spite of the deceptions they have caused me, that I should resign myself to sacrificing them."

In this final paper Poincaré conjectured, without proof, the following result: *An area preserving map of an annulus that moves the points of the boundary circles in opposite directions has at least two fixed points.* To explain this statement, we take an *annulus A*; this is the region of the plane between two concentric circles C_1 and C_2: the flat ring shown in figure 2.1. Next we define a *map* from A to itself. As in our discussion of Poincaré maps in chapter 1, this is a rule that assigns to each point of A another unique point of A. For example, it might assign to the point a, in figure 2.1, the point b. We can say that a is moved to b by the map, or that the *image* of a under the map is b. It may also occur that a point c is assigned to itself, in which case we say that c is a *fixed point*, since it does not move under the action of the map. Every point in A is associated with an image point under the map. Poincaré assumed that the map is *continuous*, meaning that the images of nearby points are likewise near one another. One should think of the annulus as a sheet of rubber, which the map distorts without tearing.

Two further conditions are required of this map. First, it should be *area preserving*, i.e., the image of any region is another region having the same area as the initial one. In figure 2.1, for instance, the region X is moved to the region Y, the areas of which are equal. This must hold for every region in A. Preservation of area may sound restrictive, but examples of

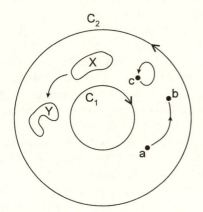

Figure 2.1. Birkhoff's fixed point theorem.

such maps abound in classical and celestial mechanics, for this property is related to the conservation of physical quantities like energy and momentum. We have already seen the important role it played in Poincaré's prize memoir. The second condition is that points of C_1 are moved around C_1 in one direction (let us say clockwise), while points of C_2 are moved around C_2 in the opposite direction (counterclockwise), as the arrows in figure 2.1 indicate: in both cases boundary points remain on the boundary. Under these hypotheses, which say little directly about what the map actually does to points inside the annulus, Poincaré claimed that there are at least *two fixed points*.

In fact Poincaré knew how to show that, if the mapping had one fixed point, then it necessarily has a second, but he was unable to demonstrate the existence of one fixed point to start with. This seemingly abstract problem was closely related to celestial mechanics. The fixed point property, if true, would have enabled him to establish that there were infinitely many periodic orbits in the three-body problem. He was studying what we now call a Poincaré map induced by the differential equation describing that problem (see chapter 1).

Poincaré was especially interested in the subject of periodic orbits. They offered an indirect approach to the problem of stability, as well as a means to understand nearby orbits. He considered this issue of utmost importance and returned to it throughout his life. In *Les méthodes nouvelles* he made the following comment: "It seems that this fact could not be of any practical interest whatsoever. Indeed, the probability is zero for the initial conditions of a motion to correspond precisely to those of a periodic solution. But it may happen that they differ by very little, and this takes place exactly in those cases where the old methods no longer apply. The periodic solution can then be advantageously taken as a first approximation, or to use Gyldén's language, as an intermediary orbit." In this context of periodic solutions, the conjecture on fixed points of an annulus map was a natural step along the way, for as we pointed out in chapter 1, fixed points of Poincaré maps correspond to periodic orbits of the underlying differential equation.

Poincaré died in July 1912, and his final paper reached Princeton that same summer. In a few months Birkhoff was able not only to understand the ideas in the paper but also to supply a proof of Poincaré's conjecture. At the end of October, Birkhoff presented a communication to the American Mathematical Society entitled "Proof of Poincaré's Geometric Theorem." The result is known today as the Poincaré-Birkhoff fixed point theorem.

Birkhoff returned to Poincaré's ideas and papers many times; in particular, he revisited the homoclinic tangle. In his book *Dynamical Systems*, published in 1927, he proved the existence of infinitely many periodic orbits, in addition to the infinite sets of homoclinic points we discussed in chapter 1. His arguments are subtle and not easy to follow. Rather than presenting them here, we shall move forward in time to the second half of the century to describe a lovely idea that enormously simplified the study of tangles and made possible a much deeper understanding of Poincaré's vision of chaos.

ON THE BEACH AT RIO

There is no Nobel Prize in mathematics. Alfred Nobel, the inventor of dynamite, was a practical man with a taste for science and literature; in fact, he wrote fiction himself. He evidently had less respect for pure mathematics, preferring to reward achievements in more directly applicable fields. The rumors that Gösta Mittag-Leffler had an affair with Nobel's wife, leading to the exclusion of mathematics from the prize subjects, are without foundation, for Nobel never married. There is, however, some evidence of personal animosity between the two men.

The foremost distinction in mathematics is the Fields Medal, created by the Canadian mathematician John Charles Fields, a former honorary president of the International Mathematical Union and professor at the University of Toronto, who formed a close friendship with Mittag-Leffler during a visit to Europe. It is via Fields that we derive our information on the mathematician and the Swedish industrialist. J. L. Synge, a colleague of Fields at Toronto (and nephew of the Irish playwright J. M. Synge), played an important role in establishing the awards after Fields's death in 1932. Regarding the exclusion of mathematics from the Nobel prizes, he wrote forty years later: "I should insert here something that Fields told me and which I later verified in Sweden, namely, that Nobel hated the mathematician Mittag-Leffler and decided that mathematics would not be one of the domains in which the Nobel prizes would be available."

It is unusual for a prize to originate within the field it honors; generally the endowments for such awards come from sources far away, as in the case of Nobel. In this connection it is not coincidental that the financial award accompanying a Fields Medal is modest, particularly in comparison to Nobel prizes. The medal has been awarded every four years since 1936, to individuals under the age of forty, at the International Congress of Mathematicians. It is widely believed that, if a researcher has not created exceptional mathematical results before age forty, he is unlikely to do so afterwards. There are, of course, exceptions to this: Karl Weierstrass, one of the renovators of mathematical analysis, whom we met as a member of the prize jury in chapter 1, spent fourteen years as a school teacher. He was recognized by the mathematical community only during the second half of his long life.

Plate 2.2. The Fields Medal.

In 1966 Stephen Smale of the University of California at Berkeley was one of the four Fields medalists announced at the Congress in Moscow. He received the prize for proving one of Poincaré's topological conjectures. Topology, to which Poincaré made great contributions, is a branch of mathematics dealing with geometrical objects that remain invariant under *homeomorphisms*. What does this mean? Let us take a geometrical object, a circle, for example. Topology does not make distinctions among circles, ellipses, or any other *simple closed curves*. ("Simple" means that the curve should not intersect itself.) All these represent one and the same topological object. We are allowed to stretch, shrink, bend, or deform objects, but usually not to cut or paste them together. If a cut is allowed, in the case of the circle, for example, it is only to make a *knot* that cannot be obtained otherwise, and with the condition that the cut ends be pasted back together. From the topological point of view, a circle, an ellipse, and a knotted string with ends joined, are all "circles." (See fig. 2.2.)

The topologist is like an ant: crawling on the circle, he sees only the local structure and cannot appreciate how the whole is embedded into three-dimensional space. But a circle, which has no ends, and a line segment, which has two, *are* recognized as distinct topological entities. Similarly, a two-dimensional sphere is different from a two-dimensional plane, as we noted in our discussion of manifolds in chapter 1. One can define spheres of any dimension. Poincaré's conjecture, too technical to describe here, involved n-dimensional spheres, and Smale proved it for all dimensions greater than or equal to five. The four-dimensional case was settled only recently, and that in three dimensions is still unproven. It is the premier unsettled conjecture in topology, if not in all of mathematics.

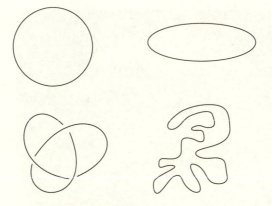

Figure 2.2. Four topological circles.

Now our story concerns dynamics, and we mentioned the sphere conjecture only to illustrate how different areas of mathematics may come together in one mind. In fact, when Smale received the award, his interests were no longer in topology per se. He had switched to the theory of dynamical systems sometime earlier, and was working on iterations of maps. Smale considered this theory important not only for the role that first return maps play in the study of differential equations, but also because the study of maps provides a parallel theory to that of differential equations, easier in some respects. He hoped to transfer results from one to the other. More generally, his earlier, purely topological work gave him a new and powerful perspective on long-standing problems in dynamics.

In late December 1959 Smale had left his temporary position at Princeton for Brazil, with his wife Clara and their two children: Laura, two and a half, and Nat, six months. He had been invited by Mauricio Peixoto to spend part of his second year's postdoctoral fellowship at IMPA, the Institute of Pure and Applied Mathematics in Rio de Janeiro. Peixoto's work on structural stability (chap. 4) was not at that time widely known and the Institute was a rather low-key affair, housed in a small old building in the suburb of Botafogo. (This would change in the years ahead, particularly when Jacob Palis, who was to study with Smale, returned to Brazil.)

Once the family was settled in Rio, in the apartment of an air force officer who had fled after an abortive coup, Smale began to spend mornings on the beaches with a pad of paper and a pen. He frequented the Leme beach, next to the far more famous one of Copacabana. Elon Lima, a friend from IMPA, would sometimes join him. Most people cannot work in public places, but Smale found he could concentrate well. A swim now and then and the breeze from the Atlantic kept him fresh.

Plate 2.3. Stephen Smale. (Photograph: G. Paul Bishop, Jr.)

He was not yet familiar with Poincaré's *Les méthodes nouvelles* and did not know that he was to be drawn into the homoclinic tangles described at the end of that book. For the moment he was working on a particular differential equation. Sometime earlier, he had formulated a conjecture concerning the structure of fixed points, periodic orbits and their stable and unstable manifolds for *structurally stable* or "typical" differential equations and mappings. (We shall discuss structural stability in chapter 4.) In particular, he felt that such systems should possess only finite sets of periodic orbits in any bounded region of their phase spaces. If there were infinitely many periodic points in a finite region, some of them must be packed arbitrarily closely. Surely any small change in the map would destroy such a delicate structure. Indeed, Peixoto had already shown this to be true for differential equations with two-dimensional phase spaces, like those pictured in figures 1.1–2, 1.6, and 1.9–10.

At this point, Norman Levinson, of the Massachusetts Institute of Technology, suggested that Smale look at the *van der Pol equation*, which describes the oscillations of an electrical circuit subjected to a periodically varying input voltage. Named after a Dutch physicist working at the Phillips Electrical Laboratories, it is similar in some respects to the equation modeling the periodically perturbed pendulum, which, as we saw in chapter 1, can have chaotic solutions.

Mary Cartwright and John Littlewood of Cambridge University, working during the Second World War on the development of radar, had shown that van der Pol's equation can have infinitely many coexisting periodic solutions as well as what Birkhoff had called "discontinuous recurrent" solutions. There was some interplay of physics and mathematics in their approach; in fact, in the paper describing the mathematical result, they note that their belief in it was for a time sustained only by experimental evidence from electrical circuits, due to van der Pol and his colleague van der Mark. Levinson had simplified and studied the problem further in a paper published in 1949 in the *Annals of Mathematics*.

Now this is a particular equation and it might be very special; Smale's conjecture could still hold true for "most" cases, but Levinson thought it might lead to a counterexample. Actually, had Smale known of Poincaré's work, he would have already had his counterexample. But the research literature is so vast: if one read all the relevant papers, no time would remain for original research. Scientists must often rely on casual conversations and hints dropped at conferences to point them in the right direction. In any case, Smale would come to know *Les méthodes nouvelles*, as well as Birkhoff's work, very well over the next few years.

Trained as a topologist, Smale was used to understanding the relations among mathematical objects through pictures. His intuition was stimulated by drawings, which he would subsequently translate into rigorous analytical language. Levinson's paper, which in essence considers the Poincaré map of the van der Pol equation but contains almost no pictures, challenged him to draw strips, to shrink, to stretch, and to bend, and then to intersect or overlap them as if in a children's game. It was not easy; still, he was happily engrossed.

Such a mood in the midst of difficult work is rather typical for a scientist. Research requires great energy, time, and dedication. It is impossible to pursue it for long without enthusiasm. When the curiosity and joy in discovery is gone, there is little remaining. Mathematics has the advantage of providing an inexhaustible source of problems that can be followed in sequence, forming a bridge toward a permanent, unreachable goal. It is perhaps the goal as much as achievements along the way which keeps one happy. From this point of view there is a close resemblance between the feelings of a dedicated mathematician and a deeply religious believer.

SMALE'S HORSESHOE[*]

    Smale focused on a particular part of the Poincaré map of van der Pol's equation. He tried to idealize and simplify it. After the first iteration of his map he had a picture that looked like a curl superimposed on a square. The square represented a set of initial conditions for the equation. The curl was the corresponding states one cycle later: their images under the Poincaré map. In the summer of 1960, when he spoke on the subject at Berkeley, Lee Neuwirth asked, "Why don't you make it look like this?" and drew a picture on the blackboard. Smale complied. Neuwirth's picture was a little simpler and more appealing than the one that occurred naturally in van der Pol's equation. Smale was primarily interested in general properties of differential equations, so he could afford to make such esthetic changes. He first shrunk the square horizontally, then stretched it vertically and finally bent it and replaced it to intersect the original square, as shown in figure 2.3. The shape suggested the name *horseshoe*. The second iteration of the map again implied, in geometrical terms, a shrink, then a stretch, and finally a bend, but now of the horseshoe-shaped image. Smale obtained the picture of figure 2.4. Toward the end of this chapter (fig. 2.9), we will show how his map arises in the neighborhood of *any* transverse homoclinic point.

  In this and the two following sections we will explore the dynamics implicit in the horseshoe map in some detail. As in similarly technical parts of chapter 1, some readers may find our arguments and constructions hard to follow, but we encourage them to keep trying, and to resort to paper and pencil when pure visualization fails. Those electing to skim these parts can pick up the story in "Oscillations and Revolutions."

Smale was interested in the set of points that would forever remain in the square. It would include all the periodic solutions as well as the recurrent solutions mentioned above. To obtain it, he had to remove from the square all those points whose images under the map lay outside the square, not only after the first iteration, but after *any iteration, forwards or backwards*. In the first iteration, the part of the square that gets bent into the "arch" falls outside, so he focused on the other parts whose images overlap or *intersect* the square. In this way he avoided the difficult analysis of points in the curved arch and could concentrate on the simpler stretching and shrinking behavior. As we will see, the consequences of this are rich enough.

  Imagining that the square is a thin sheet of rubber that we can pick up, bend, and replace, we see that the intersection of its first image with the original square consists of two vertical strips, as shown in figure 2.3. The intersection of the second image with the square yields four thinner

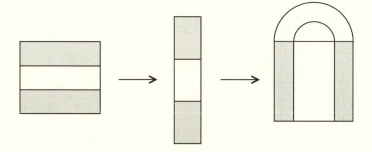

Figure 2.3. The process of shrinking, stretching, and bending for the first iteration of the map describing Smale's horseshoe.

vertical strips, which are in turn contained in the previous two strips. (See fig. 2.4.)

Continuing this process, Smale obtained at the third iteration a new, more complicated image that intersects the initial square in eight vertical strips. All eight are included, in pairs, in the previous four and consequently in the two initial ones. Iterating this scenario, one doubles the number of strips at each step, every new strip being thinner than those of earlier generations. After four steps there are 16 strips (2 raised to the power 4), after five steps, 32, and after 20 steps, 1,048,576 (2 raised to the power 20). After infinitely many steps, what remains is a so-called *Cantor set* of vertical line segments. The initial square has been stretched vertically and shrunk horizontally to an infinite thread, which is coiled back down on itself infinitely many times. We cannot draw this, no matter how fine our pencil point, but we can imagine it and deduce its properties.

This peculiar set is named after Georg Cantor, a nineteenth-century German mathematician who invented it in his quest to understand the notion of correspondence between infinite sets. We will have more to say about set

Figure 2.4. The second iteration of the map gives rise to a bent horseshoe.

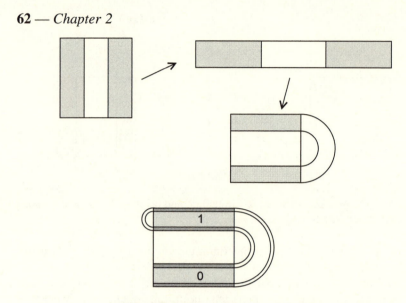

Figure 2.5. Inverse iteration of the map.

theory a little later, and Cantor sets will reappear in chapter 3. For years they were thought to be purely abstract constructions, but now, via a Poincaré map and Smale's topological ingenuity, a Cantor set was emerging naturally from a differential equation.

The infinite set of vertical line segments comprises all points that are still in the square after forward iterations of the map. Where did they originate in the initial square? To answer this we must consider inverse images, obtained by running time backwards in the differential equation. Now strips are stretched horizontally and shrunk vertically, so one obtains figure 2.5. The *pre-image* of the Cantor set of vertical line segments is evidently a Cantor set of horizontal line segments. If one picks an initial condition in such a segment, then its future images will remain trapped in the square.

We noted that Smale was concerned with points that remain *forever* in the square, which implies they are trapped under both forward and backward iteration. As figure 2.5 shows, the middle parts of the horizontal line segments are mapped outside the square on the first inverse iteration. We must now remove them, together with all other points that fall outside after any inverse iteration. But this information is already implicit in figure 2.4! We have noted that the Cantor set of vertical lines comprises those points that are still in the square after arbitrarily many forward iterations, so they are exactly the points that never stray under *inverse* iteration. Evidently, the set of points in the initial square that never leave under forward *or* backward iteration is that belonging *both* to the vertical and horizontal Cantor sets. We shall call this set Λ.

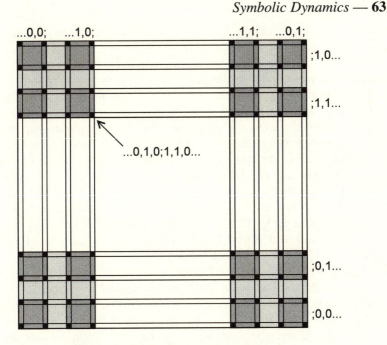

Figure 2.6. Toward the cloud of points, Λ.

It is perhaps easiest to imagine Λ by first drawing the regions that remain trapped for one forward and one backward iteration. This is the intersection of the thickest vertical and horizontal strips, shaded delicately in figures 2.4 and 2.5, so it is four squares lying in the corners of the original square. The set trapped for two iterations forwards and two backwards is gotten by intersecting the thinner strips of these figures, yielding sixteen smaller squares contained, in fours, in the four squares of the first stage. These are shown darker in figure 2.6. At each further iteration, the number of squares increases by a factor of four and their sizes shrink until, in the infinite limit, we have a cloud of points. We illustrate the first three stages in figure 2.6, the 64 smallest squares being shown darkest.

We have found that a deceptively simple map, bending a square into a horseshoe and laying it back on the square, creates an infinite Cantor set of trapped points. If this is not wonderful enough, now something really remarkable occurs. Smale realized that he could attach a code or *address* to every point in Λ, which would not only describe its location, but also tell its whole history and future under iteration of the map.

Let us denote the two big horizontal strips in figure 2.5 by the symbols 0 and 1. Suppose that x is any point in Λ. We determine its address as an infinite sequence of 0's and 1's by the following rule: if after k iterations the image of x lies in the strip 0, we write a 0 in the kth place; if it lies in the strip 1, we write a 1 in that place. It must lie in one or the other strip at each step, for if it did not, it would leave the square on the next iteration. This also goes for inverse iterations: if the point lay in strip 1 five iterations in the past, we put a 1 in the −5th place. So to find where our point will have reached after twenty iterations, or where it was thirty iterations ago, we simply look at the appropriate entry in the sequence. Thinking of our eyes as fixed and the sequence written frame by frame on a film strip, we have to wind the strip on or back, as if viewing a movie. Each (forward) iteration of the map simply shifts the whole sequence one place to the left; inverse iterations move it to the right. Conventionally, we locate the origin or center of the sequence by putting a semicolon directly before the zeroth entry, so that if a_k denotes the kth entry, the central part of a sequence is written $(\ldots, a_{-3}, a_{-2}, a_{-1}; a_0, a_1, a_2, \ldots)$. This *bi-infinite* sequence is necessary since the maps and differential equations we are concerned with run equally well forwards and backwards.

The location of x is deduced from its sequence by considering the behavior of forward and backward iterations, illustrated in figures 2.4 and 2.5. For example, the point whose sequence has central block $(\ldots, 0, 1, 0; 1, 1, 0, \ldots)$ lies in the upper strip 1, since its zeroth entry is 1. Its first iteration likewise lies in the strip 1, which, from figure 2.4, implies that the point itself must lie in the lower substrip of this upper strip, for after stretching, the map turns this strip upside down. But its second iteration lies in the lower strip 0, implying that, one step earlier (after the first iteration) it lay in the *upper* part of 1. Unwinding the effect of the map once more, we see that the point belongs to the fifth of the thinnest horizontal strips in figure 2.6, counting upward. Similar reasoning reveals that it lies in the fourth of the thinnest vertical strips from the left. The corresponding square, one of 64 coded by all possible central sequences of length 6, is indicated on the same figure. The best way to understand how the various symbolic addresses are located is to spend some time with paper and colored pencils.

Thus Smale connected his horseshoe map with the *shift map* σ acting on a set of infinite symbol sequences. (The idea of shifts goes back to the Bernoullis, whom we met briefly in chapter 1.) In fact, Smale showed that there is a perfect correspondence between the horseshoe map and the shift map: any property of the shift map is true for the horseshoe map, and conversely. In this way he reduced the study of the horseshoe to that of shifts on symbol sequences. As we shall see, this was not just a curious coincidence, but represented real progress.

Mathematicians often make advances by showing that the problem they are trying to solve is secretly the same as another problem that someone else has already solved. (This is not really cheating: there is often a lot of work in demonstrating the correspondence!) In this case it was very useful, for the properties of the shift map were well known. They belong to a body of theory called *symbolic dynamics*. But this was not the only reason to celebrate. A key property of the shift map is its *chaotic* character. We will shortly explain what this means and then give a precise definition of chaos. It is important to do this here, for, over the past ten or twenty years, the notion of chaos has been abused and misused not only by nonspecialists, but even by scientists inside and outside mathematics. Many irregular or disordered phenomena, just because they are not steady, or periodic, or otherwise nice and easy to understand, have been described as chaotic. Although we are allowed to call a thing whatever we like, in bestowing the same name on several hundred different objects we create nothing but confusion.

In new fields, where progress is rapid and disorganized, it is difficult to settle on the appropriate terminology quickly and efficiently. It is not always clear which axioms and definitions will lead to the most useful results. Nevertheless, the main reason *chaos* is abused is probably not due to this; it rather lies in the intriguing and exotic connotations of the name. "Chaos theory" has been characterized as a "new science" and excessive claims have been made for it. Similar claims were made in the case of *catastrophe theory* until it found its more modest place and value in the sciences. While the name *chaos* remains exotic today, it has begun to lose some of its less well founded connotations. Now that funding agencies are no longer scrambling to support chaotic proposals, it might be a good time to sit down and rethink the terminology. It seems best to leave this, along with the careful laying of foundations, to mathematicians, while physicists and other applied scientists keep them on their toes by testing their new definitions and results in practical applications. It is in this way that *calculus* and *probability theory*, used today as tools by so many scientists outside mathematics, were developed. A true *chaos theory* might follow a similar route. Until then, we prefer the less loaded, if more awkward name: *dynamical systems theory*.

SHIFTS ON SYMBOLS*

Symbolic dynamics was born around the beginning of this century in the work of Jacques Hadamard. It was further developed in the thirties by Marston Morse and Gustav Arnold Hedlund, who applied it to the *calculus of variations* and *differential geometry*. Symbolic dynamics has been enormously important in the development of dynamical systems theory, as in other fields of mathematics. Our presentation of the subject will be restricted here to those properties of the shift map directly connected with Smale's horseshoe, although it may at first seem that we have abandoned the world of dynamics. For the moment, we ask the reader to approach symbolic dynamics as a kind of mathematical game. We will relate it to horseshoes and chaos later.

Denote by Σ the set of all possible infinite sequences formed by 0s and 1s. The first question is, "How does one measure the distance between two elements of Σ?" Suppose that $(\ldots, s_{-2}, s_{-1}; s_0, s_1, s_2, \ldots, s_n, \ldots)$ and $(\ldots, r_{-2}, r_{-1}; r_0, r_1, r_2, \ldots, r_n, \ldots)$ are two such sequences, in each of which every element s_n and r_n, for all integers n, takes either the value 0 or the value 1. The two sequences are equal if and only if $\ldots, s_{-2} = r_{-2}, s_{-1} = r_{-1}; s_0 = r_0, s_1 = r_1, s_2 = r_2, \ldots, s_n = r_n, \ldots$, meaning that *all* corresponding entries are equal. Two sequences are considered close if they agree on a central block of symbols ($s_j = r_j$ for all j between $-N$ and N). The longer the coincident block of entries (the larger N), the closer the sequences. For example, the sequences

$$(\ldots 0, 1, 0, 1, 1; 1, 0, 1, 0, 0, \ldots)$$

and

$$(\ldots 0, 1, 0, 1, 1; 1, 0, 1, 1, 0, \ldots)$$

are closer to each other than the sequences

$$(\ldots 0, 1, 0, 0, 0; 1, 1, 1, 1, 1, \ldots)$$

and

$$(\ldots 0, 1, 1, 0, 0; 1, 1, 0, 1, 1, \ldots).$$

In the former pair, the central six entries agree, while in the latter, only the central four entries coincide.

The next question is, "What is a periodic sequence?" These are sequences in which a finite block of elements repeats itself forever. For example, the sequence $(\ldots, 1, 1, 0, 1, 1, 0; 1, 1, 0, 1, 1, 0, \ldots)$, in which every pair of 1's is followed by a 0, is a periodic sequence. After three shifts to the left, the sequence returns to its former self. Here the period is 3. Similarly, the sequence $(\ldots, 0, 0, 0, 1, 1; 0, 0, 0, 1, 1, 0, 0, 0, 1, 1, \ldots)$ has period 5. In this way one can construct infinitely many periodic sequences in the set Σ.

It is easy to see that the shift map σ acting on a periodic sequence gives rise to a periodic orbit. Take, for example, the sequence

$$S_0 = (\ldots, 1, 0, 1, 0; 1, 0, 1, 0, 1, 0, \ldots)$$

and apply the shift map σ to it. The first iteration gives the sequence

$$S_1 = (\ldots, 1, 0, 1, 0, 1; 0, 1, 0, 1, 0, \ldots)$$

(recall that we shift to the left, which is the same as moving the central semicolon one space to the right). The second iteration of σ yields (. . . , 1, 0, 1, 0, 1, 0; 1, 0, 1, 0, . . .), or S_0 again, the third gives again S_1, and so on. The sequences S_0 and S_1 alternate at each iteration of σ, giving rise to a periodic orbit. In this way we have also discovered what a *periodic orbit* of the shift map σ means. Instead of the analogy of the movie, one might think of a slide projector with a circular carrousel, so that after showing one's neighbor's vacation pictures all the way through, it returns to the start and repeats. More generally, the orbit of any sequence under the shift map is the collection of all sequences obtained by shifting the original sequence as many steps to the left or right as one wishes. For periodic sequences, this is a finite set, like that composed of S_0 and S_1, but as we shall see, for "most" sequences it is infinite. The pictures never repeat.

To get a foretaste of how such a simple symbolic machine can encompass chaos, suppose that we toss a coin repeatedly, noting *heads* or *tails* at each toss. Identifying the head as 0 and the tail as 1, we produce a random sequence of 0's and 1's. But this sequence, and any random sequence produced in any like manner, also belongs to Σ. So, along with all the periodic sequences, Σ contains many sequences which might as well have been chosen at random. Hence Σ has orbits that never repeat, but dance about chaotically.

The last notion we must clarify is that of *density*. This is not at all easy to imagine. We start with a geometric example. Think of a curve C drawn on a sheet of paper P. Typically there is plenty of "space" around C and so many points in P that are not close to C, but in fact one can (theoretically) construct a curve that visits near *all* points in the plane of the paper. In this case we say that C is dense in P. Somehow C must wind about almost all over P. (Can you imagine how this might happen? In chapter 5 we will give examples of such curves on annuli and tori.) Similarly, an orbit of the shift map σ will be called *dense in the set* Σ, if for *any* given sequence in Σ the orbit contains a point that lies arbitrarily close to the given sequence. Following this dense orbit, one comes as close to *every* element of Σ as one wishes.

Set theory is a branch of mathematics dealing with collections of objects or *elements*, called *sets*, and which endeavors to establish truths about sets, regardless of the nature of the elements composing them. Set theory acquired a bad name when it was imposed on innocent children during the "New Math" debacle. However, a set theoretical formulation of density, which generalizes the properties sketched above, will reveal other aspects of this notion, and so a brief foray into this abstract world is worthwhile.

An orbit of σ and the curve C can be viewed as sets contained in the larger sets Σ and P, respectively. Roughly speaking, a set is dense in some other set if the elements of the former approximate very well those of the latter. For any element of the second set, one can find an element of the first that lies arbitrarily close to it. A classical example is the set of *rational numbers*, which lies within the set of all *real numbers*, or just *reals*. The set of rationals consists of those numbers that can be written as fractions: 1/3, 7/4, 2/5, etc. But the rationals do not account for all possible real numbers. The remaining reals are the set of irrationals, formed of those numbers that cannot be written as fractions: $\sqrt{2}$, $\sqrt{3} - 1$, π, etc. (It was a great triumph of Greek mathematics to realize that such numbers exist.) The fact that the set of rationals is dense in the set of the reals means that *any* real number can be approximated as closely as we want by a rational. For example $\sqrt{2}$ can be estimated by 1.41 = 141/100, but it is better approximated by 1.4142 = 14,142/10,000, and even better approached by 1.414214 = 1,414,214/ 1,000,000. This sequence of approximations can be continued indefinitely, and each further estimate improves on the previous one. The rational numbers are therefore dense in the reals. On the real number line we can picture the convergence of a sequence of rational approximants to an irrational as in fig. 2.7.

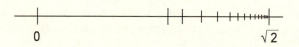

0 $\sqrt{2}$

Figure 2.7. Rational numbers approximating $\sqrt{2}$.

SYMBOLS FOR CHAOS[*]

We can now reenter the world of *chaos* with a safety net of mathematical rigor beneath us. Although there is still no unanimously accepted mathematical definition of a chaotic map, most specialists consider *chaos* to be characterized by the following properties:

1. Sensitive dependence on initial conditions,
2. Existence of at least one dense orbit,
3. Density of the set of periodic orbits.

These properties are not independent, in certain cases (1) with (2) implies (3).

We have listed the properties in order of their practical importance, but it is easier to understand their meaning in converse order. Starting with the density of the set of periodic orbits, we shall show that the shift map exhibits each of these features. Using the definition of distances between sequences given earlier, we can easily create periodic orbits (= sequences) that are as close as we wish to any given sequence. Suppose that sequence is $S = (\ldots, s_{-3}, s_{-2}, s_{-1}; s_0, s_1, s_2, \ldots)$ and consider the periodic sequence $T = (\ldots, s_{-1}, s_0, s_{-1}; s_0, s_{-1}, s_0, \ldots)$. The middle two entries of these sequences agree. A better approximation is provided by $R = (\ldots, s_0, s_1, s_{-2}, s_{-1}; s_0, s_1, s_{-2}, s_{-1}, \ldots)$, which agrees with the given sequence on four central entries. The idea is now obvious: one makes a periodic sequence by copying a block of length $2n$ from the middle of the sequence one wishes to approximate, and then repeating that block ad infinitum in both directions. The bigger n, the better the approximation. There is no limit to how large we may take n, as long as it is finite. This demonstrates the density property (3) for the shift map.

Next we establish the existence of a single dense orbit. We will construct a sequence such that the corresponding orbit of the shift map contains points that are close to *every* element of Σ. Define the symbolic sequence S^* by

$$S^* = (\ldots; \underbrace{0, 1,}_{\text{one-blocks}} \underbrace{0, 0, 0, 1, 1, 0, 1, 1,}_{\text{two-blocks}} \underbrace{0, 0, 0, 0, 0, 1, \ldots,}_{\text{three-blocks}} \ldots).$$

S^* is constructed as follows. First, we write all possible blocks of length 1, which simply means 0 and 1. We next write all blocks of length 2: 0, 0, then 0, 1, followed by 1, 0 and 1, 1. We continue this process with three-blocks, four-blocks, and so on, forever, tacking all the finite sequences end to end. The resulting infinite sequence still has a beginning. To make it infinite in both directions, we just reverse a copy of it and paste the two together in the middle. We have built a sequence in which all possible (finite) subsequences of every length occur.

A little imagination shows that this is the sequence we are looking for. Indeed, since iteration of the shift map is equivalent to sliding the sequence along so that distant parts come into focus, and S^* contains all finite sequences, *some* iteration of σ applied to S^* will produce a sequence that agrees with any given sequence on as many central entries as we wish. This proves that, for any given sequence in Σ, there is an iteration of S^* under σ that matches the given sequence as closely as we like. We have shown that property (2) holds.

We have finally arrived at the most important characteristic or symptom of chaotic behavior: sensitivity with respect to the initial data. This is, in the eyes of many, the kernel of chaos. What does it mean? Remember the continuity of solutions of differential equations with respect to the initial data, described in chapter 1. It states that orbits starting close together will stay close together *for a while*. All the systems we are concerned with obey this condition. We also introduced what may seem a similar notion: that of stability, which demands that orbits starting close together stay nearby *forever*. Sensitive dependence on initial conditions does not contradict conti-

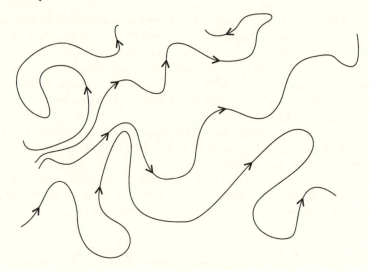

Figure 2.8. Sensitive dependence with respect to the initial data.

nuity, but it is at total odds with the stability property. Only for *short* intervals of time can we guarantee that orbits starting together, stay together. To ensure that the close encounter lasts longer, we must start the solutions with yet closer initial conditions.

But if sensitive dependence prevails, for almost all orbits this interval will be quite short. After some time, the solutions will separate and henceforth behave in quite different manners. In figure 2.8 some curves have close encounters, after which they diverge rapidly. Recalling figure 1.6c, it is as if the phase space were sprinkled with saddle points whose unstable separatrices drive the neighboring orbits apart. In fact, this is precisely the case for Smale's horseshoe: the infinitely many periodic orbits are all of saddle type.

The property of sensitivity of solutions with respect to initial conditions was understood, at least in part, by James Clerk Maxwell. The famous Scottish mathematician and physicist wrote an essay in 1873 in which he pointed out that when microscopic perturbations give rise to macroscopic changes, the prediction of future events is effectively impossible. Sensitive dependence is a mathematically precise way to express the highly unstable and unrepeatable character of a system's behavior. Its implications are far-reaching. In fact, Maxwell's remarks were delivered at Cambridge University as part of a larger debate concerning free will and determinism. Poincaré would echo these ideas some thirty years later in his essay "Science and Method."

We can experience this property in our everyday lives. Some people know that if they leave their houses at a certain time in the morning, the highway will be only moderately busy and they will reach their workplace in good time. But if they drive off just five minutes later than usual, the heavy traffic will delay them for an additional half hour. A slight change in the initial conditions leads to a large variation in the journey time.

Another—by now classical—example of sensitive dependence has suggested the name "butterfly effect." In 1972 the meteorologist Edward Lorenz addressed the American Association for the Advancement of Science on "Predictability: Does the Flap of a Butterfly's Wings in Brazil Set Off a Tornado in Texas?" Of course, he was not suggesting that the tiny force of the butterfly could directly *cause* a tornado. Rather, he saw the inherent instability of the atmosphere (or more precisely, of the equations modeling it) as amplifying the small cause into a large effect.

It is not difficult to see that the shift map has the sensitive dependence property. Suppose two sequences, S and R, are close together. This means their central $2n$ entries coincide, for some large number n. Yet, after n iterations of σ, applied to both S and R, we might obtain two sequences far apart from each other, since their $(n + 1)$st (and $-(n + 1)$st) entries may differ. For example, the central four entries of $S = (\ldots, 0, 1, 0, 1; 1, 0, 0, 1, 1, \ldots)$ and $R = (\ldots, 0, 1, 0, 1; 1, 0, 1, 0, 0, \ldots)$ coincide. After two iterations we have $(\ldots, 0, 1, 1, 0; 0, 1, 1, \ldots)$ for S and $(\ldots, 0, 1, 1, 0; 1, 0, 0, \ldots)$ for R. The shifted sequences differ in the 0th place. S and R are close, but two iterations of σ already make the sequences separate. This establishes property (1): sensitive dependence of the shift map with respect to the initial data. We have proved that the shift map is chaotic in the precise sense of our definition.

We noted earlier that Smale demonstrated a "perfect correspondence" between the horseshoe map acting on points in the Cantor set Λ and the shift map acting on sequences in Σ. As in topology, the perfect correspondence is a *homeomorphism*, a continuous, one-to-one transformation. Smale had to show that each element of Σ correponds to just one point in Λ and vice versa. This was done first by taking finite sequences of length—say, six— showing that each one corresponds to one of the $2^6 = 64$ little squares of figure 2.6, as we argued above. He then let the length n of the sequences increase and noted that the corresponding squares shrink to points as n approaches infinity. To prove continuity, he had to show that nearby points in Λ correspond to nearby sequences in Σ. But these properties are implicit in the very definition of a sequence as the "address" or location of a point: all sequences with central block $(\ldots, 0, 1, 0; 1, 1, 0, \ldots)$, for example, lie within the square marked by this sequence in figure 2.6; all sequences with central block $(\ldots, 1, 0, 1, 0; 1, 1, 0, 1, \ldots)$ lie in one of the four yet smaller squares inside that one, and so on.

One of Stephen Smale's big contributions to the theory of dynamical systems was this connection between the horseshoe and the shift map. Similarly, a link can be made between the Poincaré map, near a transversal

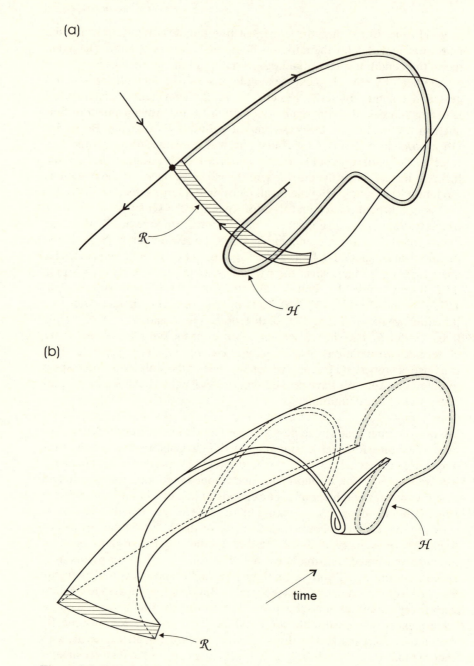

(a)

$\mathcal{R}$

$\mathcal{H}$

(b)

$\mathcal{H}$

time

$\mathcal{R}$

Figure 2.9. Transverse homoclinic points imply horseshoes: (a) in the map, (b) in the differential equation.

homoclinic orbit, and the horseshoe map. Figure 2.9 shows how, after sufficiently many iterations of the map of figure 1.14, a thin rectangle $\mathcal{R}$ is stretched and shrunk and bent into the horseshoe shape $\mathcal{H}$. It reveals that the horseshoe map, the Poincaré map near a homoclinic orbit, and the shift map are, in essence, only different ways of describing the same phenomenon: *chaos*. In figure 2.9 we also sketch how the horseshoe appears in the three-dimensional phase space of the differential equation as the flow squeezes and stretches, like a baker kneading dough. The human mind needed almost three quarters of a century to link these ideas and to become comfortable in switching among the analytic, the geometric, and the symbolic formulations. From a mathematical viewpoint, the easiest one to understand and handle is undoubtedly the shift map. In fact, it is so easy to deduce properties such as those we have described above, and the descriptions are so clear that the name *chaos* seems unwarranted. This illustrates the power of symbolic dynamics and of mathematics in general, in which superficially different ideas are linked and shown to be equivalent.

Thus, the problem of the homoclinic tangle, with which Poincaré had wrestled at the end of the last century, continued to reveal its secrets through the results that Smale obtained in the 1960s. Recent formulations of the complete result describe it as the *theorem of Poincaré-Birkhoff-Smale*. It has become a cornerstone in the study of nonlinear dynamics.

As a footnote to this story, it is interesting to note that, after the sixty-year lapse between Poincaré's work and Smale's, there was a further delay before publication. Although Smale essentially worked out the horseshoe and its symbolic dynamics during his stay at IMPA in 1960, he did not get around to writing the first paper describing it until late 1961. He had first to finish other papers in differential topology, his earlier interest. The paper itself, with the modest title "Diffeomorphisms with Many Periodic Points," did not appear until 1965, and then in a volume dedicated to differential and combinatorial topology, probably read by very few dynamicists.

OSCILLATIONS AND REVOLUTIONS

Between the times of Poincaré and Smale, the foremost mathematician in the field of dynamical systems was George Birkhoff. After moving to Harvard, he went on to attack yet more difficult problems than the one on fixed points. He became an expert on Poincaré's research in dynamics, and worked to understand the most delicate points left unsolved by his predecessor. Among these were the intriguing phenomena occurring near the transversal intersection of the stable and unstable manifolds of a fixed point for the first return map. As we described in chapter 1, Poincaré realized that this led to a homoclinic tangle, but as he admitted in *Les méthodes nouvelles*, he felt unable to draw it or describe its properties.

In 1927 Birkhoff published his masterpiece, *Dynamical Systems*. As we have remarked, in this book he proved that, in any neighborhood of a transverse homoclinic point, there are infinitely many periodic orbits. In the symbolic description developed by Smale, these correspond only to *part* of the infinite set of periodic sequences. Certain papers and books, including a text co-authored by one of the present writers, therefore refer to the homoclinic tangle result as the *Smale-Birkhoff theorem*. But to leave out the name of Poincaré is like cutting the roots of a tree and praising the fruit alone. Some newer references call the result the "theorem of Poincaré-Smale." As we have suggested, it seems fairer to include all three names.

Birkhoff actually made a greater contribution to this problem than most dynamical systems theorists realize. In 1935 he published another book entitled *Nouvelles recherches sur les systèmes dynamiques*. In the fourth chapter he studied further consequences of the transversal intersection of invariant manifolds. However, this book was somehow overshadowed by the earlier one, and few people read it. Smale actually rediscovered and formalized many of the results contained in this later work of Birkhoff.

This fact does not at all diminish Stephen Smale's remarkable insights into the subject. Equipped with the right choice of formalism and the "language" of symbolic dynamics, Smale was able to go much further than his predecessor. The discovery of the *horseshoe* is primarily due to Smale, for although some of the ideas behind it, and attempts to find a symbolic language, can be recognized in Birkhoff's papers, the latter lacked the right (mathematical) words to describe it completely.

Symbolic dynamics is much more generally useful than this particular application reveals. We will now give an example of how the technique is used directly in the three-body problem, which was, after all, the original stimulus for Poincaré. For a long time, researchers in celestial mechanics sought an example of a so-called *oscillatory* solution, i.e., an orbit that almost escapes to infinity but continually returns. It is easy to construct a function like this; the graph of one is shown in figure 2.10. We just take oscillations that are amplified at each step. Since the function takes small values as well as arbitrarily large ones again and again, it has no limit as time increases to infinity. Colloquially speaking, it cannot decide what it wants to do in the end. Of course, one can draw whatever one likes and the question remains: Are there orbits of the three-body problem that have such wild behavior?

This question may seem foolish, but its answer would prove important in the classification of solutions of the *n*-body problem with respect to their

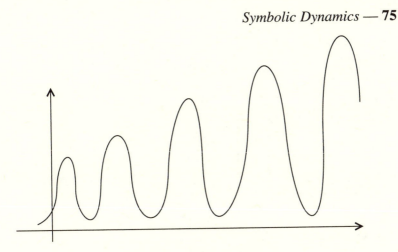

Figure 2.10. An oscillatory function.

asymptotic behavior. In the next chapter we will probe this issue in greater depth. Every science aims to classify and to organize its knowledge, and mathematics makes no exception. The asymptotic properties of solutions of a differential equation (i.e., the behavior of systems after a long, long time) can be tackled only theoretically. No computer is capable of following orbits for infinite times, and so pure thought must take us there. Celestial mechanicians wanted to know if such oscillatory beasts lived among the other, tamer sorts of solutions they had already found.

In 1961 the Russian mathematician K. Sitnikov proved that such a solution does exist. He probably learned of the problem from A. N. Kolmogorov, whom we shall meet in chapter 5, although Sitnikov was working in topology at the time and was not actually Kolmogorov's student. He considered two particles, *A* and *B*, of equal masses, called the *primaries*, which move in ellipses around each other, in a plane, following the laws of Kepler. He took a third particle, *C*, of very small mass, so small that its gavitational field does not affect the motions of *A* and *B*. Initial conditions were chosen such that *C* remains for all time restricted to the line ℓ, perpendicular to the plane of the primaries and passing through their center of mass, as shown in figure 2.11. The horizontal components of the gravitational pulls of *A* and *B* on *C* exactly cancel in this case, so that *C* can only shuttle up and down the vertical line. This is another variant of the *restricted three-body problem*, different from the one on which Poincaré had worked (where all three bodies lie in the same plane; see fig. 1.8).

Sitnikov showed that, for suitable initial conditions, the motion is such that the particle *C* moves first above and then below the plane of the primaries, increasing its distance with each oscillation. At each pass through the plane, its energy receives a small boost from the motions of *A* and *B*, so that

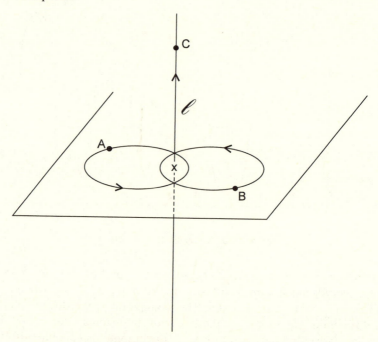

Figure 2.11. Sitnikov's problem.

it can travel farther before being drawn back. Throughout, the primaries move unperturbed on fixed ellipses, providing an infinite reservoir of energy that allows the excursions of C to grow without bound. Nevertheless, C comes back every time, and so the motion has no limit in infinite time.

A few years later another Russian mathematician, V. M. Alekseev, a student of Kolmogorov, returned to Sitnikov's example and, using symbolic dynamics, showed that *any kind* of solution can appear in this problem, depending on how one chooses the initial conditions. Moreover, he proved that this remains true even for a particle C of finite mass. Here *any kind* is understood in the sense that the set of solutions forms a one-to-one correspondence with the set of infinite sequences of symbols consisting of *any* natural number, not just 0 and 1 as in the horseshoe. How is this done?

The idea is simple. Count how many rotations the primaries perform between two consecutive passages of C through the plane containing A and B. Then construct a bi-infinite sequence by assigning to each element of the sequence the corresponding number. For example take the sequence $(\ldots s_{-2}, s_{-1}; s_0, s_1, s_2, \ldots) = (\ldots 5, 3; 2, 7, 1 \ldots)$. Now 2 lies in the place s_0, so there are two rotations of the primaries up to the first passage of C through the plane. Since 7 corresponds to s_1, there will be seven rotations between the first and the second passage of C through the plane. In the same way we find one rotation between the second and the third passage, and so on. We proceed similarly in the past, for the negative entries, finding that there were three rotations immediately before the zeroth passage through the plane. Since the orbital periods of A and B do not change, the fact that we can put any integers in the sequence implies that the amplitude of the oscillations can be made arbitrarily high. The difficulty lies in establishing the correspondence between rotations of the primaries, passages of C through the plane, and the symbol sequences. It is very remarkable that, for *any* sequence we choose, there is a corresponding orbit of C.

This brief story illustrates the involvement of two Russian mathematicians in a major problem in celestial mechanics. It is interesting to note that, especially since the Second World War, Soviet mathematics has followed a path of its own, producing many results of great value, but largely in isolation. The cold war and Soviet politics, on one hand, and the language barrier and lack of ready translations, on the other, created for several decades an almost impassable divide between the Western and Soviet scientific cultures. Soviet scientific journals reaching the West often contained few details and only partial results; Western journals were unavailable in many Soviet institutions. While mathematics developed extensively in both regions, the lack of communication tended to create parallel universes. Many Western researchers have been unpleasantly surprised to find, when the news filtered through, that some of their results had been discovered, sometimes years in advance, by Soviet mathematicians. Of course, the converse was also true.

Numerous Western mathematicians tried to prevent this waste of energy, traveling to the Soviet Union and Eastern Bloc countries any time an opportunity arose and they were fortunate enough to obtain a visa. Smale was one of those who traveled to Moscow to make connections between the two worlds. We will meet another such visitor in chapter 5. Reciprocal visits were much harder to arrange, being only occasionally permitted. After some mathematicians from behind the iron curtain used these opportunities to defect, travel restrictions grew even tighter.

Today, following the disintegration of the Soviet Union and the "revolutions" in the former Eastern Bloc countries, everything has changed. But as in the larger spheres of politics and economics, not all these changes have been beneficial. Many members of the younger generation of Eastern mathematicians have immigrated to the West, incidentally making the job market for young Westerners highly competitive. More fundamentally, there has been a clear shift of talent toward Western Europe and North America. There is now a serious gap in the former Communist countries, which have the added problem of finding financial support for their universities and research institutes.

These developments were impossible to foresee during the period when Sitnikov, Alekseev, and Smale were obtaining their results. The blind competition of the years that followed will probably never occur again. Mathematical developments might have been very different had this rivalry been more open. In our final chapter, we will see how Khrushchev's "thaw," and the slight opening to the West that followed, had a remarkable effect on the work of at least one Soviet mathematician: A. N. Kolmogorov. The more dramatic revolutions of the late 1980s and early 1990s will probably have yet more influential consequences.

Symbolic dynamics is used extensively today. It is hard to imagine the development of research in dynamical systems without this valuable technique, and it has found applications in many other branches of mathematics and computer science as well. In the latter it is used in coding theory and in designing algorithms for data storage and transmission. Symbolic methods have also become important instruments in numerous other areas of applied science and engineering, especially now that the prevalence of chaos has become clear. In 1991, for example, a conference on symbolic dynamics was held at Yale University in honor of Roy Adler of the IBM Research Laboratories. It brought together 139 participants from thirteen countries, representing not only pure and applied mathematicians, but also physicists, computer scientists, and electrical engineers from universities and industry.

A New Science?

We have made a trip into the heart of chaos, a *new science*, as some popular accounts would have it. But is it really a new science? The history we have told runs counter to this assertion. Chaos was discovered by Poincaré, partially explored by Birkhoff, and better understood by Smale. Born in the 1880s, it is scarcely new. The many other achievements of the last two decades, even when obtained independently, are in a sense subordinate to those of these founders. All the various descriptions and representations of chaos ultimately reduce either to symbolic dynamics or to one of the finer criteria derived by mathematicians more recently. Anything beyond these is probably speculative and unlikely to have much scientific value. Neither the vague notion of "irregular," nor the precise one of "unstable," mean chaotic.

In fact we doubt that chaos can properly be called a *science* at all. According to the 1992 edition of *Webster's College Dictionary*, a science is "a branch of knowledge or study dealing with a body of facts or truths systematically arranged and showing the operation of general laws." While the techniques developed for studying chaos may satisfy this definition, they are a part of an already well-established science: mathematics. More precisely, they are a part of the mathematical theory of dynamical systems, and not such a large part at that. In no way do they constitute a new science in the sense that ecology or computer science do. The attribution of this term seems largely literary hyperbole.

What about the more modest claim for a chaos theory? Alas, even here we feel there is some misstatement. A *theory* typically addresses a limited range of phenomena, providing laws or models with which they may be described. Relativity theory and quantum theory, the cornerstones of modern physics, are cases in point. In contrast, chaos encompasses a loose collection of *methods* for the *analysis* of differential equations and other dynamical systems. Symbolic dynamics, Poincaré maps, and the other tools we have described do not tell us anything *directly* about the world; rather they help us understand and analyze the consequences of *other* theories and models. One might say that chaos fails to be a theory because it applies to too much.

Nonetheless, the recognition that chaos exists and is even common in our deterministic models of the world *has* changed our thinking, and from this point of view it has perhaps done more than a theory or even a new science. The broadly accepted view that chaos is common in our deterministic models of the world has been a revolutionary step. It has led to a fresh philosophy, a new way of looking at phenomena that evolve over time or even space. It influences almost every scientific and technological field, stimulating researchers to invent new methods and concepts. It has brought a new spice to literature, the arts, and culture in general, and even to the rarefied realms of critical theory. It touches almost every human activity, from the launch of satellites to television commercials and popular books and movies such as *Jurassic Park*. Whether we like it or not, it has become part of our lives.

The old paradigm of cycles, of regularity and periodicity, has not been replaced—it has acquired a new interpretation. It can be seen now as a first approximation to a richer and more complex vision of the universe. The ancestral myth of eternal return can never be told again in quite the same way. At best we can retain the idea of recurrence (in the sense of Poincaré's ergodic theorem, to be discussed in chapter 4), but we should not hope for exact repetition. This is perhaps a reasonable compromise between order and chaos.

3.

Collisions and Other Singularities

> Painlevé's intellectual activities were characterized by
> an exceptional breadth, force, and efficiency in the ser-
> vice of an almost prophetic mind. He required only a
> few years to succeed, in the most remarkable way, in
> mathematics, mechanics, and politics.
> —René Garnier

THE CHAMBERLAIN entered the lecture hall and
announced: "His Majesty, the King!" The audience stood, its soft murmur
transformed to silence. All eyes looked toward the door. No one moved.

The heavy steps of a tall, gray-haired man broke the silence, and the
entire room burst into applause. King Oscar II of Sweden and Norway was
entering one of the large auditoriums at the University of Stockholm. A
remarkable event: the king was attending an introductory lecture at one of
his country's most prestigious academic institutions.

How many presidents, prime ministers, or heads of state—in or out of
power—have paid attention to a public presentation at some university?
(We are not thinking here of political campaigning or occasions of high
academic politics, such as the conferral of honorary degrees on themselves
or others, but of routine lectures for students.) It is hard to name even one.
Had His Royal Highness come today for such an ordinary event? In one
sense *yes*; in another *no*. One motive behind the king's presence was to be
found in his own personality; the other lay in the identity of the speaker: the
talk to follow was more than a *routine* lecture.

Oscar II of Sweden and Norway was one of Scandinavia's most remarkable rulers. Educated by the historian F. F. Carlson, Oscar succeeded to the
throne in 1872 on the death of his older brother Charles XV. He was a fine
orator, a poet, and a patron of the arts and sciences. His deep respect for
culture, gained probably through his former teacher, was the main reason
for his establishment of prizes such as that won by Poincaré, as we described in chapter 1. In fact, Oscar had provided financial backing in 1882
for the founding of the journal *Acta Mathematica*, in which Poincaré's
paper had appeared, as well as issuing awards to individual mathematicians. The king also made a habit of inviting to Stockholm, from time to

Plate 3.1. Paul Painlevé. (Courtesy of Centre Nationale de la Recherche Scientifique, Paris)

time, some of the more prominent artists and scientists of Europe. On that day, 2 October 1895, King Oscar was to hear the first in a series of lectures offered by a visiting mathematician.

As the waves of applause subsided, the chairman began his introductions. After the customary royal greetings, he introduced his distinguished guest, "Professor Paul Prudent Painlevé of the University of Paris!" The room erupted for a second time in applause as a young man walked to the podium and turned to face the audience. In the fall of 1895, Painlevé was not quite thirty-two years of age. He was known to be an excellent speaker. His courses at the Faculté des Sciences were often oversubscribed. His erudition was combined with a natural and straightforward manner, even in the formal context of a public lecture. It was a true pleasure to listen to him.

"Your Highness," he began, "since Your Majesty has the kindness to honor this lecture with your presence, my first duty is to express my respectful gratitude to Your Majesty for the great honor of being called to Stockholm to discuss recent progress in Analysis. . . . As Your Majesty has indicated, the main object of this course is the theory of transcendents, and in particular the uniform transcendents defined by differential equations. I would like to describe the central place occupied by this theory in the research of these last years. This will allow me in some measure to justify the spirit and direction of modern mathematics."

The introductory formalities had been fulfilled, and now Painlevé entered the world of science. He began with Newton and Leibniz, going on to discuss Kepler's laws and showing how the idea of differentiation led to the gravitational law and the *n*-body problem. Finally, he outlined the work of Bernoulli, Euler, Clairaut, d'Alembert, and others who had also been pioneers in the field of differential equations.

In the huge, packed lecture hall, few among the audience had much knowledge of these topics. In spite of this, Painlevé's balanced, elegant style engaged all his listeners. He was easy to understand. He presented the essence of difficult material without tedious mathematical detail. There was not a whisper in the room, no breath; the audience was striving to absorb each word and enjoy every sentence. They were enthralled.

A thunder of applause filled the auditorium as he finished. The king himself showed his admiration: proud of the decision to invite Painlevé to Stockholm, he stood up and smiled contentedly while signaling his approval. The opening had been a brilliant success, as were the twenty-two lectures to follow.

A Singular Man

As the second sentence of Painlevé's first lecture stated, the goal of his course was to study so-called *transcendents* defined by differential equations. To understand this rather technical notion, we must first explain *singularities* of solutions to differential equations. For this, we shall return to the metaphor of flowing water, as in chapter 1. We suggested there that the vector field defined by a differential equation resembles a river current that carries pieces of driftwood (individual solutions) along on its surface. Sometimes a stick or branch may be swept to shore and beached—the analogue in the vector field is that at some specific time the solution leaves the phase space: it literally ceases to exist. Somewhat colloquially we say that *it blows up in finite time*, for often it ceases to exist by becoming arbitrarily large: the solution is carried off to infinity. Alternatively, it may meet an obstacle in phase space. In our example of the falling ball in figure 1.1, the surface of the ground at $h = 0$ is a physical obstacle. Only positive heights (positions) are permitted, so the phase space is not the whole plane, but merely the right-hand half-plane, corresponding to *all* velocities (positive and negative) and to all positive

(a)

(b)

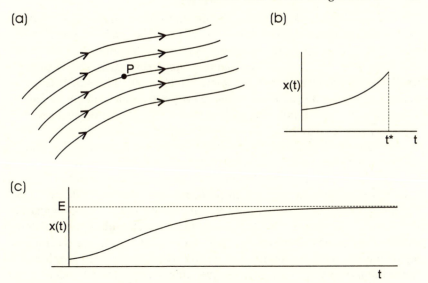

(c)

Figure 3.1. The orbit C stops at point P. On the orbit C, the point P corresponds to the singularity t^*. (a) Shows the phase portrait, and (b) the singular solution as a function of time. In contrast, (c) shows a different solution approaching an equilibrium point.

heights. The line $h = 0$ is the "river bank" in this case; when the ball reaches it on the way down, the solution ceases to be defined. (We could add a model to describe how the ball bounces off the ground. However, this would involve issues such as the ball's elasticity and impact dynamics, and would take us beyond our "simple" Newtonian world.)

Singularities may also occur at isolated points in phase space, as sketched in figure 3.1a. In general, if a solution fails to be defined for all values of the time variable t and instead stops at some finite instant t^*, we call t^* a *singularity* of the corresponding solution.

Before going further, we remark on a crucial difference between a solution having a singularity and one tending to an equilibrium point of the flow, as defined in chapter 1. Figures 3.1a and 1.6 suggest that both solutions end at a particular point, but what these phase-space pictures fail to show explicitly is that the former stops at a finite time while the second takes an infinite amount of time to reach its limit at the equilibrium. One needs a graph of the solution as a function of time, as in figures 3.1b and c, to appreciate the difference. In one case the curve stops at time t^*; in the other it is defined for all time.

To explain what a transcendent meant to Painlevé, we first point out that "illusory" singularities may occur. In our metaphor, the stick may drift to shore, only to be immediately swept back into the current due to the particular shape of the bank. Mathematicians call such a singularity *regularizable*, meaning that the solution may be extended beyond the time t^*. In this case we say the singularity is merely *apparent*. The stick goes on to float down the river again, until it encounters another obstacle: a new singularity. When a singularity is not apparent, Painlevé called it *transcendent*. By definition, a solution cannot be extended beyond a transcendent singularity t^*. In the n-body problem, singularities occur when collisions between two or more particles take place, as we shall see in a moment. If the motions of the colliding particles can be extended in a meaningful way beyond the collision, the singularity is apparent; otherwise it is transcendent.

In his Stockholm lectures, the French mathematician presented the difficult subject of singularities and in particular that of transcendents. Things appear simple in the analogy with the river, but for differential equations, singularities can cause a lot of trouble. Painlevé classified many different cases, provided new methods for their study, and found numerous interesting results. As a central application he considered the n-body problem. The first natural question to ask was whether singularities actually occur in this classical problem, and, if so, what is their physical interpretation?

An obvious example of a singularity in the n-body problem is that of a *binary collision*. At the instant t^*, two particles coincide, sharing the same position coordinates. The set of points in phase space having two equal coordinates is a *hyperplane*. In particular, when the phase space is the plane, the hyperplane is just a straight line, such as $h = 0$ in figure 1.1a. This simple case occurs only in a rectilinear central force problem: one body moving toward or away from a center on a fixed line. As for the falling ball, the two coordinates of phase space are the position and velocity of the moving particle with respect to the center. Since the force in Newton's gravitational attraction law is inversely proportional to the square of the distance, and the distance between the particle and the center is zero at collision, the force becomes infinite and the equations describing the motion no longer make sense. Therefore, the line having the zero position coordinate that defines the collision is an obstacle for orbits in a planar phase space. When an orbit reaches it, there is no obvious way of continuing beyond or glancing off the line (see fig. 3.2).

The same thing occurs in binary collisions between any two particles, and in triple collisions, quadruple collisions, etc. *Simultaneous collisions* also cause obstacles in phase space. An example of this kind is a *double binary collision* in which two particles collide at one point in physical space while, simultaneously, two other particles collide somewhere else. Although these are very unlikely events, they must be understood if we are to have a *complete description of the n-body problem*. Moreover, like the

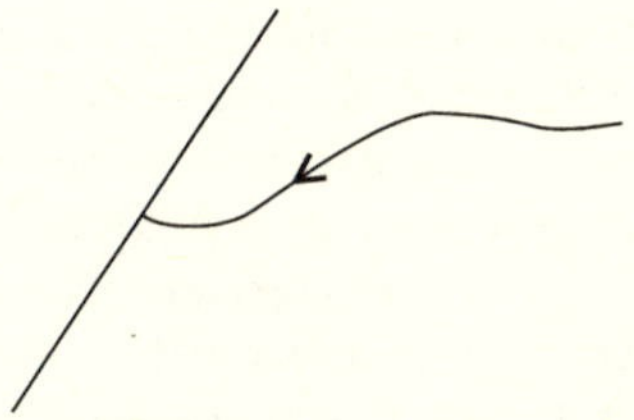

Figure 3.2. An orbit reaching a hyperplane that defines the collision in a rectilinear central force problem.

stone thrown in a pond which sends out ripples, collisions exert their influence over the structure of phase space in a whole neighborhood of the singular points.

We have answered, in part, the questions raised above. We have given examples of singularities, and we know their physical interpretation. But, as usual in science, each answer prompts new questions. We must now ask if collisions are transcendents or only apparent singularities. We may also wonder whether collisions are the only possible singularities in the *n*-body problem. These are difficult problems, and Painlevé was unable completely to solve them in his lectures. He gave a partial answer to the second question, proving that, *in the three-body problem, the only singularities are collisions*. He attempted to extend this result to more than three bodies, but failed. At this point he stated a conjecture.

Conjectures stimulate the development of mathematics. In their research, mathematicians sometimes feel intuitively that they have the correct answer to a question, but in spite of prolonged attempts cannot prove it. It is common at this point to formulate a *conjecture*, i.e., to state clearly what the right answer to the question should be, in the opinion of the author. The longer a conjecture endures without proof or disproof, the more famous it may become. Conjectures are especially important if key results in the evolution of a field depend on their clarification. Attempts to clarify such problems may encourage the growth of a whole branch of mathematics or lead to the creation of others. We have already mentioned Poincaré's conjecture on *n*-dimensional spheres, remarking that the three-dimensional case remains the major unsettled conjecture in topology.

Perhaps the most famous conjecture in all of mathematics is Fermat's Last Theorem, named in honor of Pierre de Fermat, a French lawyer, who wrote it in the margin of a book (and claimed to have proved it) over three hundred years ago. Fermat claimed that no (nonzero) integers *a*, *b*, and *c* can be found that satisfy the equation $a^n + b^n = c^n$, when *n* is an integer greater than 2. In 1993 the mathematical world was stirred by the announcement that Andrew Wiles of Princeton University had proved Fermat's theorem, but in the painstaking process of review and checking, a gap emerged. In 1994 Wiles and his former student Richard Taylor succeeded in closing the gap. In his eight-year struggle with the problem, Wiles introduced important new ideas that go far beyond the special case of Fermat's claim. In doing so, he has brought geometry and number theory closer together. This illustrates another value of conjectures: they draw attention to and concentrate talent on particular areas, yet the light cast in their solution often extends beyond the immediate problem.

Painlevé's conjecture concerns singularities other than collisions. It states that *for n larger than three, there exist solutions with singularities that are not due to collisions.* On the penultimate page, number 588 of his meticulous handwritten manuscript for the Stockholm lectures, Painlevé sketched in a single sentence his idea of how such a solution might appear. The particles would oscillate wildly, come close to a collision, scatter apart before coming yet closer to another collision, and this scenario would repeat again and again in a finite period of time. He was, however, unable to prove that such a complicated dance could actually occur in the *n*-body problem.

Painlevé said no more about the motion, adding only that Poincaré had previously mentioned such possible *pseudocollisions*; it is not clear in what context or where, since Painlevé gives no reference. Perhaps the idea was never published but merely arose in conversation between the two Parisian mathematicians. Nevertheless, today the conjecture bears the name of Painlevé and it has stimulated, in the intervening hundred years, the many achievements that we shall explore in this chapter.

Shortly after the Stockholm lectures, Paul Painlevé began to move beyond his mathematical activities. In 1897, at the time of the Dreyfus case, he entered politics and, without abandoning mathematics, reached the highest levels of French government. Elected deputy of the fifth *arrondissement* of Paris and reelected in 1914, he was made minister of public instruction in 1915–16 in the government of Aristide Briand. On 20 March 1917, Painlevé received the war portfolio, and that September he became prime minister. Due to a conflict with left-wing circles, he had to step down in November of the same year. As president of the Chamber of Deputies in 1924, after the resignation of Alexandre Millerand, Painlevé ran for the presidency of France. He was not elected, but became prime minister again from April to November 1925. In October and November he also held the finance ministry. Between 1926 and 1929 Painlevé regained the war portfolios under the governments of Aristide Briand and Raymond Poincaré (a younger cousin of Henri). Cambridge University awarded him the title of *Doctor Honoris Causa* in 1927. He was elected deputy again in 1928 and

points du système tendent vers des positions limites à distance finie, ou bien il existe au moins quatre points du système, soit $M_1, \ldots M_\mu$ ($\mu \geqslant 4$) qui ne tendent vers aucune position limite à distance finie, et qui de plus sont tels que le minimum $\varphi(t)$ de leurs distances mutuelles tende vers zéro avec $t-t_1$, sans qu'aucune de ces distances tende constamment vers zéro.

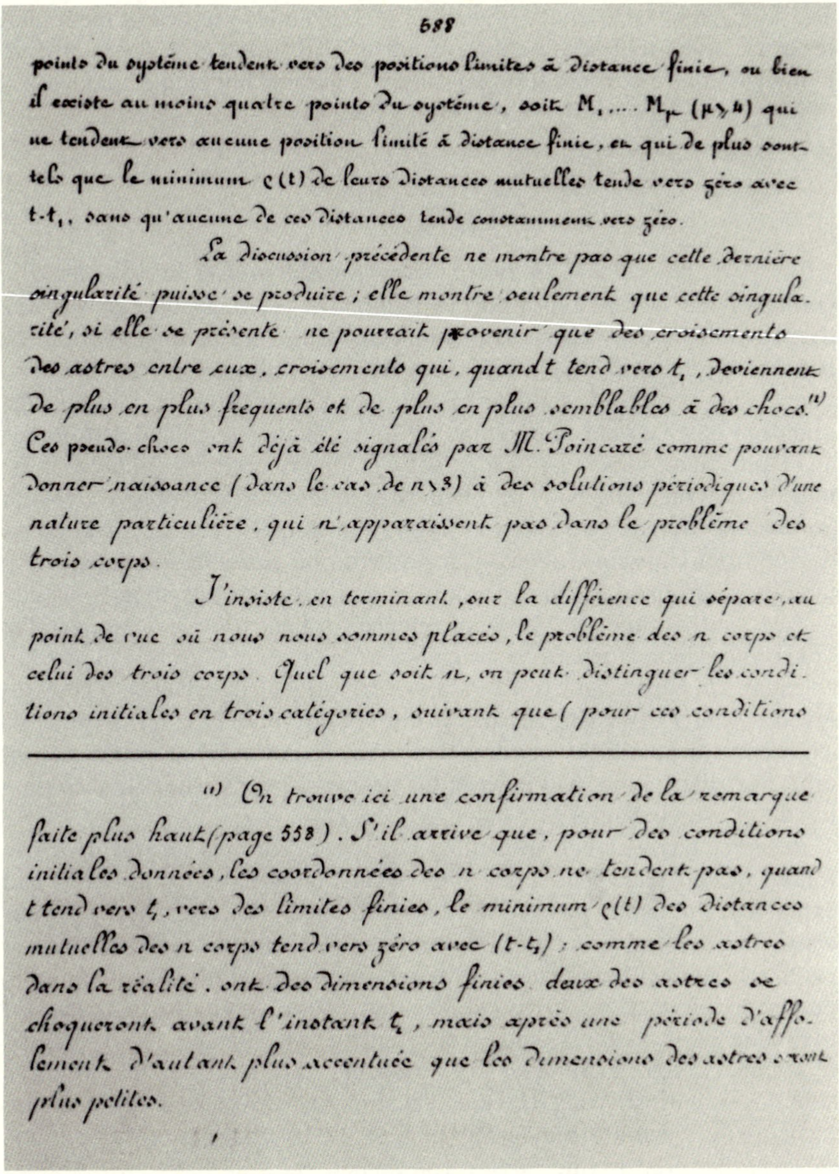

La discussion précédente ne montre pas que cette dernière singularité puisse se produire ; elle montre seulement que cette singularité, si elle se présente ne pourrait provenir que des croisements des astres entre eux, croisements qui, quand t tend vers t_1, deviennent de plus en plus fréquents et de plus en plus semblables à des chocs.[1] Ces pseudo-chocs ont déjà été signalés par M. Poincaré comme pouvant donner naissance (dans le cas de $n \geqslant 3$) à des solutions périodiques d'une nature particulière, qui n'apparaissent pas dans le problème des trois corps.

J'insiste en terminant, sur la différence qui sépare, au point de vue où nous nous sommes placés, le problème des n corps et celui des trois corps. Quel que soit n, on peut distinguer les conditions initiales en trois catégories, suivant que (pour ces conditions

---

[1] On trouve ici une confirmation de la remarque faite plus haut (page 558). S'il arrive que, pour des conditions initiales données, les coordonnées des n corps ne tendent pas, quand t tend vers t_1, vers des limites finies, le minimum $\varphi(t)$ des distances mutuelles des n corps tend vers zéro avec $(t-t_1)$; comme les astres dans la réalité ont des dimensions finies, deux des astres se choqueront avant l'instant t_1, mais après une période d'affolement d'autant plus accentuée que les dimensions des astres sont plus petites.

Plate 3.2. Painlevé's conjecture, from the manuscript of his Stockholm lectures. (Courtesy of Centre Nationale de la Recherche Scientifique, Paris)

1932, and from 1930 until his death in 1933, Painlevé served as the minister of air under three different governments. His tomb is in the Panthéon, along with those of other former political, military, and intellectual leaders of France.

COLLISION OR BLOWUP

A twenty-two-year-old student was sitting in a corner of the lecture room during Painlevé's opening address. He was deeply impressed by all he saw and heard. Although he understood only parts of the lecture, it was to have a decisive influence on his life. The young man's name was Edvard Hugo von Zeipel. Born in Sweden, the grandson of a German immigrant, he loved mathematics and astronomy, and Painlevé's course was a happy combination of both. It is likely that the idea of studying the *n*-body problem came to him during Painlevé's visit to Stockholm. Some years later, in 1904, von Zeipel received his doctoral degree from Uppsala University for a thesis on periodic solutions of the three-body problem. Then a rare opportunity arose: he was invited to study in Paris, at that time the Mecca of science. In June 1904 he arrived in the flourishing capital, where he was able to take courses from Poincaré in celestial mechanics and from Painlevé in rational mechanics. These revived his interest in singularities of the *n*-body problem. He became intrigued by the possibility of noncollision singularities. What would such an orbit look like? What had been in his teachers' minds when they conjectured this wild behavior?

Research begins when something is unclear. For most of our lives, we follow beaten paths and learn only what others have experienced and explained. But nature has endowed us with a priceless characteristic, curiosity, an ancestral drive toward the unknown. In two million years of evolution, human beings have developed this natural gift into the highly organized intellectual activity that we now call research. We have learned to find methods, to apply them in different areas of life and, more recently, of science; to fight to understand every detail. It is, however, the primitive stimulus of curiosity that provides the strength to overcome difficult moments in our search for understanding. In the first decade of our century, this was the force driving Edvard von Zeipel.

In September 1906 von Zeipel left Paris and returned to Sweden. He came home rich with scientific experience and ready to tackle difficult problems. He wanted to go beyond Poincaré and Painlevé and was determined to follow up on the singularity question. Within two years of his return, in May 1908, he published a paper, in French, with the modest title, "On the Singularities of the *n*-Body Problem." In barely four pages,

Plate 3.3. Hugo von Zeipel. (Courtesy of Uppsala University Library)

the article stated and proved that *a necessary condition for a solution to have a noncollision singularity is that the motion of the system becomes unbounded in finite time.* Roughly speaking, his result implied that, if a solution is singular, it must either lead to collision, or blow up in finite time.

To make this clear, let us express it in a different way. Von Zeipel claimed that, if all mutual distances between particles remain finite, then the only possible singularities are collisions. So, for a noncollision singularity, the distance between at least two of the particles must become unbounded. It need not grow infinite by virtue of the particles separating forever; they might oscillate back and forth with the separation distance increasing at each step and becoming unbounded, rather as in Sitnikov's restricted three-body problem, described in chapter 2. But to qualify as a singularity, this must all happen in a finite interval of time. Now, von Zeipel did not prove that noncollision singularities *do* exist. He only described what must happen *if* they do. His result implied that, without the above behavior, all singularities are necessarily collisions. Advances in mathematics, as in the other sciences, usually take place in small steps like this one. Von Zeipel had established the connection between noncollision singularities and oscillatory solutions, showing that the two are inextricably linked and that it is therefore enough to study only one of these phenomena.

This result has an interesting history. In 1920, the French astronomer Jean Chazy published a paper in *Comptes rendus hebdomadaires* in which the same claim occurs without proof and without reference to von Zeipel. Evidently Chazy was unaware of the 1908 paper. The first printing in 1941 of Aurel Wintner's book, *The Analytical Foundations of Celestial Mechanics*, refers to von Zeipel's paper, claiming that the proof has gaps and casting doubt on the validity of the argument. A detailed demonstration came in 1970, when Hans Sperling, of the Marshall Space Flight Center in Huntsville, using an idea similar to von Zeipel's, overcame all the obstacles besetting the problem. It seemed that the issue was closed. However, as we described earlier, in 1985 Richard McGehee from the University of Minnesota spent part of his sabbatical leave at the Mittag-Leffler Institute near Stockholm. He found von Zeipel's paper, which had appeared in a little-known Swedish journal held in the library of the institute, and read it carefully. McGehee realized that von Zeipel's proof was in fact correct and translated it into modern mathematical language, more easily accessible to specialists today. He published a paper in 1986, establishing the truth and von Zeipel's priority.

After 1910, von Zeipel's interests drifted slowly toward more practical astronomical aspects of celestial mechanics. He wrote a paper on perturbation methods, applied his knowledge to the motion of asteroids and comets, and finally turned to astrophysical research on the structure and evolution of stars. In 1915 he was elected to the Swedish Royal Academy of Sciences and, four years later, appointed professor of astronomy at the University of Uppsala. In 1920 Mittag-Leffler asked von Zeipel to contribute to the memorial issue of *Acta Mathematica*, dedicated to the life and work of Poincaré. Von Zeipel wrote a seventy-five-page paper on the French mathematician's work in celestial mechanics. This shows how highly he was regarded, for only a select group contributed to the issue: Appell, Hadamard, Lorentz, Painlevé, Planck, and Wien. From 1926 to 1935 von Zeipel served as chairman of the Swedish Astronomical Society, and from 1931 to 1948 he presided over the National Committee for Astronomy. His career was crowned in 1930 with the Morrison Prize of the New York Academy of Sciences, for his contributions to the theory of structure and evolution of stars.

Although it was an early work and outside the main thrust of his later research, von Zeipel's theorem played a fundamental role in the history of Painlevé's conjecture. Many people believed that it provided a good argument *against* the existence of noncollision singularities. How could a "closed" system of particles acquire sufficient energy to become unbounded in finite time? (There is a significant difference here from Sitnikov's restricted three-body problem, in which the steadily orbiting primaries provide a boundless energy source, for we assumed that their motions are unaffected by the third, small body.) Still, this remained hard to resolve and, after Chazy, attention shifted away from the problem. In the sixties the subject was revived by two American mathematicians, Donald Gene Saari, today a professor at Northwestern University, and his former Ph.D. thesis supervisor, Harry Pollard, of Purdue University. At around the same time, numerical experiments were carried out by a group of people interested in a different question. We will describe these attempts before returning to the work of Pollard and Saari.

COMPUTER GAMES

The first digital computers were huge. In the 1950s and 1960s the wealthy universities in North America poured money into building and acquiring them. In comparison with today's personal computers and workstations, the monsters of those years filled whole rooms and worked thousands of times slower. But they had arrived, and many scientists thought this wonderful. Not everyone agreed: even today some conservatives oppose the machines and one can occasionally see on office doors defiant posters announcing "Computer-free Zone." At that time there was deep suspicion concerning computers, especially within departments of mathematics. A prominent instance of this is that, after John von Neumann's death, his former colleagues at the Institute for Advanced Study in Princeton kept their buildings entirely free of computers for many years. In addition to his enormous contributions to mathematics in general, von Neumann had been one of the co-creators of the first digital machines.

Nevertheless, in spite of such hard-line opposition, some enthusiasts were eager to use computers from the very beginning. In particular, astronomers and celestial mechanicians realized that computer simulations of Newton's equations and the n-body problem could take them far beyond the laborious hand calculations to which they had hitherto been restricted.

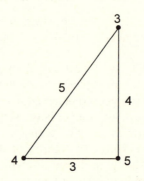

Figure 3.3. The Pythagorean problem of three bodies.

In 1966 Victor Szebehely was one of those enthusiasts. His main interests were in celestial mechanics, and the question he attacked numerically was an old one. Ernst Meissel, a former student of Jacobi, had already proposed the study of the following problem in 1893, during a discussion with Carl Burreau. (In fact, Meissel's notebooks reveal that he had done extensive computations on this and similar problems, most likely before 1893.) Consider three particles of masses proportional to three, four, and five at the vertices of a Pythagorean triangle (a right triangle having sides equal to three-, four-, and five-length units); see figure 3.3. Supposing that the bodies obey the Newtonian law of gravitation, what is their future motion if one releases them with zero initial velocities from these initial positions? The problem had been investigated numerically by Burreau in 1913 but without much success. His hand computations could not go far, so he drew no significant conclusions.

Szebehely directed Myles Standish to perform the computations at Yale University. They were continued later that year by Spinelli, Lecar, and Szebehely himself at the NASA Institute of Space Sciences in New York. Simultaneously, L. Stanek, under the direction of Eduard Stiefel, was studying the problem at ETH, the Federal Institute of Technology in Zurich. What these people observed was astonishing. After some complicated behavior, two of the particles formed a *binary* (i.e., they orbited closely around each other), while the third was expelled with high velocity. Figure 3.4 shows the results of some of these calculations.

As often happens in scientific research, the initiators of these computations were looking for something different. They were seeking periodic solutions. Instead they found the behavior we have described. To them this was even more interesting, since it shed light on the question of binary star formation. The relevance of their work to the Painlevé conjecture (in which they were not interested at the time) was that one particle is expelled, with high velocity, away from the other two. Why does this happen and why is

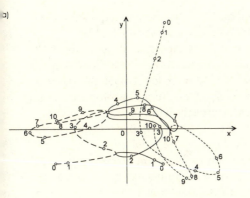

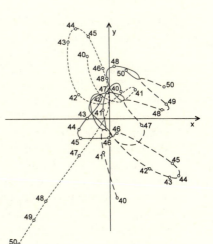

Figure 3.4. Numerical computations of the Pythagorean problem. The orbits of the three particles are represented by continuous, segmented, and dotted lines. Figures (a), (b), (c), and (d) show the motions from instants $t = 0$ to 10, 40 to 50, 50 to 60, and $t = 60$ to 70, respectively.

this property important? These turned out to be two essential questions that had to be answered in settling the noncollision singularity problem.

To answer the first question, let us take a closer look at figure 3.4b. Around the instant $t = 41$ the three particles are all very close (note that the scale in fig. 3.4b is much finer than in 3.4c or 3.4d). Under the inverse-square law of Newtonian gravity, the interparticle forces become enormous in such close encounters. So the triple approach, without a triple collision, creates a "slingshot effect," which makes the third particle accelerate strongly and rapidly leave the neighborhood of the binary.

After 1973 Jörg Waldvogel of ETH continued the numerical investigations in the general planar three-body problem. In his examples, a particle always appeared to escape. The closer the bodies came to a triple collision, the higher the velocity of the escaping one. But computer simulations cannot replace mathematical proof. Especially when velocities and forces become high, the accuracy of numerical schemes, which substitute the continuous processes of differential equations by discrete digital steps, is open to doubt. Nonetheless, these results were very suggestive. It began to look as if a closed system of orbiting bodies *could* provide enough energy to eject one particle at very high speed. Then, in 1974, in ignorance of these numerical results, Richard McGehee found an analytical proof of this property for the rectilinear three-body problem (the case when the particles only move along a fixed line). One year later, independently of McGehee, Jörg Waldvogel also published a proof for the more general case of the planar three-body problem, using a different method. Specifically, they proved that, after a close triple encounter in the three-body problem, one of the particles generally escapes from the system with arbitrarily high velocity.

The importance of this property for Painlevé's conjecture lies in the hint it provides for constructing a solution leading to a noncollision singularity. Taking four bodies, one might find initial conditions such that three of them approach in a near triple collision. Then, two will form a binary while the third is expelled at high velocity. Afterwards, the third particle must meet the fourth and, in this close encounter, be somehow flung back toward the binary formed by the first and second particles. This will lead to a new triple approach, after which the third particle is again expelled with yet higher velocity from the binary. If one can find suitable initial data, such that the triple collision approach is repeated again and again with an appreciable increase in velocity each time it occurs, the motion of the system might become unbounded in finite time. Indeed, the third particle will oscillate between the binary and the fourth particle (which themselves are moving apart), making at each stage a longer trip in a shorter interval of time. After infinitely many oscillations of this third particle, the others may have escaped to infinity in finite time, without any collision actually occurring.

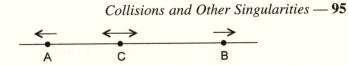

Figure 3.5. Particles *A* and *B* travel in opposite directions, while *C* oscillates between them.

This scenario may seem counterintuitive. How can infinitely many oscillations take place in a finite interval of time? Let us give an example. Suppose that two particles *A* and *B* travel in opposite directions with increasing velocities, and a third one, *C*, oscillates between these two, also with increasing velocity, such that it reaches each of the others after every oscillation (see fig. 3.5). Suppose the increase of velocity is so great that, on the first leg, *C* travels the distance from *A* to *B* in one second, on the next in half a second, on the third in a quarter of a second, on the fourth, one-eighth of a second, and so on. The total time required for *A* and *B* to reach infinity is obtained by adding all the intervals above, this means summing an infinite set of terms: $1 + 1/2 + 1/4 + 1/8 + 1/16 + \ldots$. This sum is called a *geometric series*, and one can prove that after adding all the terms the result is 2. In this case the motion becomes unbounded just two seconds from the start.

A simpler example of an infinite sequence of events taking place in a finite time is provided by the mathematical idealization of a bouncing ball. One drops the ball so that it strikes the floor after one second. Rebounding to a height somewhat less than that from which it was dropped, it next bounces after a half second, then after a quarter second, etc. After two seconds and infinitely many bounces, it is at rest. Of course, the (infinitely many) bounces in the tail of this series of events will be through distances smaller than the size of a molecule, an atom, a quark, or whatever physicists may think of next, so a real ball effectively stops moving after only a finite number of bounces.

Is it possible to construct a motion such as that of figure 3.5, making the velocities increase sufficiently at every passage from left to right and conversely? The triple approach in the *n*-body problem, for *n* larger than or equal to four, allows for such an increase. (Remember that Painlevé had shown that all singularities are collisions for $n = 3$: we can now see that this happens because there is no fourth particle to send the third one back for another encounter with the binary.) Therefore, the problem of constructing a solution that becomes unbounded in finite time is to show that the scenario described for four or more bodies can indeed take place. The only control we have is on the initial conditions. How can we be sure that initial positions and velocities can be chosen so that the third body must return to the binary over and over again? What if the close encounter with the binary or the fourth particle sends the third off in the wrong direction? Herein lies the next difficulty in proving the conjecture.

How to Catch a Rabbit

The revival of interest in the problem of non-collision singularities is due to Donald Saari and Harry Pollard. The latter was Birkhoff's "academic grandson," his thesis adviser, David Widder, having been Birkhoff's student at Harvard. Pollard was a brilliant and prolific mathematician, bursting with ideas that he knew how to exploit. In 1966 he became interested in the *asymptotic behavior* of solutions of the *n*-body problem, as time tends to infinity, i.e., in the shape of orbits after very long time. Saari, at that time Pollard's student, was seeking a subject for his dissertation. Knowing his adviser's interest, he chose to study the asymptotic behavior of solutions as they approach a singularity and enthusiastically set to work. After finishing his thesis, he and Pollard collaborated to improve some of the results. This led to the publication of two beautiful papers. It was to be the beginning of a long story.

Scientists' careers, and hence science itself, progress as much by chance as by design. It is often more a matter of being in the right place at the right time than carefully planning one's life. After completing his degree, Saari obtained a postdoctoral position in the astronomy department at Yale University. In this environment he could develop his talent, continuing to do research and publish papers. Then in 1968 he was offered an assistant professorship at Northwestern University. He moved to this post with its promise of permanence, beginning the academic work that would lead him to a distinguished career. Saari taught a course on his results on the *n*-body problem in 1970, attended, among others, by a graduate student, Carl Simon. One year later Carl accepted a postdoctoral offer from Morris Hirsch and went to Berkeley. There he used the lecture notes taken from Saari to offer a similar course. Among the audience in Berkeley was Joseph Gerver, a young man who would play an important role in elucidating the problem of noncollision singularities, but not for some years. All these figures, brought together by various chances, will reappear in our story.

The following summer, 1971, Saari gave a talk at a symposium in Salvador, Brazil, in which he introduced the noncollision singularities problem. Jürgen Moser and Richard McGehee were in the audience. As we will see later, McGehee would also make a crucial contribution in the field. And Moser, one of the greatest mathematicians of our time, was to state in his acceptance of an honorary degree in Bochum in 1990 that "the problem of noncollision singularities is one of the most important ones left in the mathematics of the twentieth century." Interest in the subject was growing.

A beautiful property proved by Saari in 1971 generalized von Zeipel's theorem. Roughly speaking, he showed that *if the configuration of parti-*

Plate 3.4. Donald Saari. (Courtesy of D. Saari)

cles in space varies slowly when time tends to the singularity, then the orbit necessarily leads to a collision. This implies that, for a noncollision singularity, the particles *must* oscillate wildly. Several people had predicted this, beginning with Painlevé; others doubted that it could happen; but nobody had been able to prove either was true. Can a system produce such wild oscillations and at the same time generate an unbounded solution in finite time? How can particles avoid colliding when they come close together? In spite of Saari's success, the problem seemed to be getting harder and harder. Questions multiplied faster than answers, but little by little some pieces of the puzzle were falling into place.

Another problem Saari was thinking about concerned those solutions free of noncollision singularities. As we noted earlier, Painlevé had proved that, in the three-body problem, all singularities are due to collisions. In 1973 Saari showed that if n particles move on a line, the first singularity they encounter is necessarily a collision. Subsequently he proved that if solutions do not come close to a multiple approach, then there is no possibility of noncollision singularities.

There are many jokes about scientists in general and especially about mathematicians. One of these is to ask, "How does a mathematician catch a rabbit?" The answer is, "He catches two, and lets one run away." In many cases this joke is not far from the truth. Some problems seem so hard that all attempts to solve them fail. One then steps back to consider a more general problem, including the hard question and maybe many others besides, and attempts to solve the general one. Surprisingly, in many situations the trick works.

Saari decided to attack singularities from a different point of view. He applied the rabbit hunt tactics: he asked how large the set of solutions with singularities is, within the set of *all* solutions of the *n*-body problem. This question came to mind after he had settled a conjecture due to Littlewood, which we will shortly describe in detail. An answer to the new problem was not supposed to explain the original one, but to give a better understanding of how many solutions with collision and noncollision singularities one might expect to find. But how does one determine the size of a set of solutions? Fortunately, the mathematics necessary for this was already in place. It is called (naturally enough) *measure theory*, and we will first present a few ideas that will help us to understand Saari's result.

We are used to measuring many things, each with a precision appropriate to the situation. We can deal with lengths, areas, volumes, as long as the objects to be measured look "nice." The problem becomes more difficult if we leave usual objects and start to think about more complicated examples. What is the length of Norway's coast? One could first compute a rough approximation, but taking the fjords into account, the result changes. Looking even closer we see that these fjords contain many smaller fjords, and so on. We can always find a better approximation to the length, going around small bays and headlands, then boulders, then stones and pebbles. . . . What is then the *real* length of Norway's coast? We realize there is no clear answer to this question and we may now wonder if the length is finite at all!

The coastline problem has found its resolution in the theory of *fractals*. It is beyond our goal to describe this subject, which has led to volumes of beautiful and strange computer pictures attempting to show behavior at finer and finer scales. We will only need to know what it means for a set to have *positive Lebesgue measure* or *zero Lebesgue measure*. The name is that of Henri Lebesgue, who defined the notion at the beginning of this century. Roughly speaking, the fact that a set M has measure zero within a space of given dimension means that if one chooses a point in this space at random, then it is infinitely improbable (although not impossible) that the point chosen belongs to M. On the contrary, if the set M has positive measure, there is a finite chance that the point belongs to M. We will give some examples of sets contained in a one-dimensional space. For simplicity, we take this to be the segment [0, 1]: all real numbers between 0 and 1, including the end points; see figure 3.6.

First take a segment contained in [0, 1], for example [1/4, 1/3]: all num-

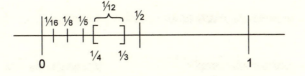

Figure 3.6. Sets in the interval [0,1].

bers from 1/4 to 1/3. Its Lebesgue measure is $1/3 - 1/4 = 1/12$, a positive number: it is just one twelfth of the total length. It can be shown that any segment that is not a pure point has positive Lebesgue measure, with value equal to the segment's length. Next take the set formed by three points {A, B, C }; this will have Lebesgue measure zero: each point has measure zero, and the sum of finitely many such also has measure zero ($3 \times 0 = 0$). Choosing a point at random, our chances of picking any prespecified point are zero. Now consider a set of infinitely many points, {1, 1/2, 1/3, 1/4, 1/5, ... }; this also has zero Lebesgue measure, although it is not so obvious. In these cases the Lebesgue measure coincides with the usual notion of length, so it may seem intuitively reasonable that a set of *points*, no matter how many there are, should have zero length.

Let us now move to a more complicated example. We shall define a *Cantor set*, similar to the set Λ described in connection with Smale's horseshoe. For this, take the segment [0, 1] and remove the middle-third: the open segment (1/3, 2/3) (i.e., the middle segment except the end points 1/3 and 2/3). The remaining set is formed by two segments, [0, 1/3] and [2/3, 1]; see figure 3.7. Now remove the middle-third open segment from each of these. The new set consists of four smaller segments, each of length 1/9. Repeating the procedure, we get eight yet smaller segments, and so on. Iterating this process indefinitely we obtain the so-called *middle-third Cantor set*. It comprises what is left after deleting all "middle-thirds" from the interval [0, 1]. If we compute the Lebesgue measure of the segments removed (all infinitely many of them), we find that their sum equals 1. (As in the oscillations problem above, this also involves summing a geometric series.) Since the Lebesgue measure of the segment [0, 1] is also 1, and we have removed from it a set of measure 1, the remaining middle-third Cantor set must have measure zero. The same thing happens if one performs the construction by removing the middle-fifth, the middle-seventh, or any other segment of fixed proportion at each step. The dust of points left behind has zero "length," although it does have finite fractal dimension.

There are also other types of Cantor sets. Suppose we first remove the middle-third as before, but then, from the two remaining segments, remove only their middle-ninths ($9 = 3^2$). From each of the four segments left we

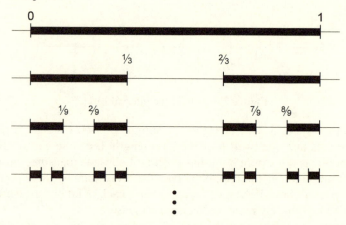

Figure 3.7. The construction of the middle-third Cantor set.

then remove the middle-twenty-seventh ($27 = 3^3$), and so on. Adding up the lengths of all segments removed, we now get a number *smaller* than 1. The Cantor set remaining at the end in this case has *positive* Lebesgue measure. Intuitively, we have taken away a successively smaller proportion at each step, so that there is more left behind. In both cases the Cantor set is a dust of disconnected points, but in the second it is a "thicker" dust. Here we have the counterintuitive situation of a collection of *points* with finite "length!" Such thick Cantor sets will reappear in connection with the Kolmogorov-Arnold-Moser theory of chapter 5.

The Lebesgue measure of a line segment in the plane is zero, but a full square in the plane—no matter how small—has positive Lebesgue measure. Though a disk in a three-dimensional space has Lebesgue measure zero, a solid sphere in space has positive Lebesgue measure. In the plane and in space there are also several exotic examples similar to Cantor sets on the segment. It is not always easy to say immediately whether a set has zero or positive Lebesgue measure.

Note that we carefully specified *Lebesgue* measure. This is because there are many other measures, bearing the names of other mathematicians. There is also an abstract *measure* that generalizes all of these. Lebesgue measure is the most common; in simple cases it coincides with our usual notions of length, area and volume.

Measure theory can be applied to all sorts of objects, not just pieces of lines, planes, and physical space. Included among them are sets of solutions of differential equations. One can ask what is the Lebesgue measure of a subset of solutions sharing a particular property, such as that of approaching a collision. We shall return to this question shortly.

A MEASURE OF SUCCESS

The spiral of the tornado was rapidly approaching. It had formed on the prairies beyond Lake Michigan, close enough to make people feel apprehensive. But this was not a rare event for Midwesterners, and few were seriously concerned. One might hear a remark like "Yeah, another tornado is predicted!" The large stone and brick buildings of the Northwestern University campus had been designed with such events in mind, probably the most turbulent ones that would occur in this peaceful place. Evanston, the host town of this respected academic institution, is a quiet, picturesque suburb of Chicago. The city with its bustling streets seems far away.

Donald Saari liked to get up before the rest of his family. His wife and two small daughters would still be asleep, so he could do some work in the early hours. Then he would eat breakfast with his family and enjoy half an hour together until it was time to kiss them good-bye and leave for his office. On his way through the park he would think of his research or of the differential geometry lecture he had to give. He enjoyed this start to the day. But that morning he could not take his usual walk. The wind was too strong, so he chose to drive to the university. Deeply engrossed in computations all morning, he entirely missed the tornado at noon. His thoughts were wrapped up in another kind of whirlwind. He had finally understood what happens in certain collision singularities. He was about to write the final sentence in a story that had occupied him for several years.

It had begun on a foggy night in the fall of 1968 when Saari was a new faculty member at Northwestern. He enjoyed a little light reading before falling asleep, and that day he had taken from the library the popular collection by J. E. Littlewood, *A Mathematician's Miscellany*. Every mathematician has probably found himself browsing through this little book at some time. It tells entertaining stories of student years and describes various events from the life of Littlewood, who, together with G. H. Hardy, dominated British mathematics in the first half of the twentieth century.

We have already mentioned Littlewood's work on the van der Pol equation, but his interests had much broader scope. One chapter in his book explains how its author had proved an interesting property of the *n*-body problem, mentioning also that the set of initial data leading to binary collisions has Lebesgue measure zero in the set of all solutions. As we have remarked, this implies that such collisions are very improbable, although not impossible. Almost all choices of initial data will lead to solutions without collisions. A footnote in that old edition, which is absent in newer ones, relates, in the manner of a joke, what a nice paradox this would create if triple collisions were probable, in the sense of taking place on a set of solutions of positive Lebesgue measure.

Saari was intrigued by the note, but did not take it seriously at the time. A few months later, however, Littlewood published a book entitled *Some Problems in Real and Complex Analysis*, in which one of the proposed open questions concerned the measure of the set of initial data leading to collisions in the *n*-body problem. After reading the statement Saari began to think about the problem again. He had an idea of how he might solve it and wanted to start at once. There was no time to waste. He knew that now that the book was published, others would also soon tackle the problem. For Donald Saari the race for the proof started that evening.

After a sleepless night, by morning he believed he had done it. He went to the university to teach his class and told some colleagues and graduate students that he had solved one of Littlewood's problems. At noon he felt exhausted and went home to take a nap. But he awoke to a big disappointment. He realized his proof was not complete. He had omitted the case of *degenerate central configurations*, a notion we will discuss in detail at the end of chapter 4. In the euphoria of discovery he believed he had found a trick to bypass the difficulty, but now he understood his computations were wrong. The problem was harder to solve than he had thought. He began to recognize some real obstacles.

The next day he told his friends that his earlier announcement was premature. He had no proof, at least not yet. But one colleague did not hear about Saari's retraction: Carl Simon. In his trips to conferences and lectures at other universities, Simon told everyone that Don Saari had a proof of Littlewood's conjecture on the *n*-body problem. After a few months Saari began to receive letters from prominent people who requested preprints of the paper—a paper that did not exist.

He was in a delicate situation. Though he continued working hard, he still lacked a full proof. In the meantime, he had abandoned his first idea and was trying other ways to solve the conjecture. Fortunately an offer from Eduard Stiefel, who invited him to spend some time at ETH in Zurich, got him off the hook, at least for a while. He packed and went to Switzerland, hoping to produce a proof during his visit and so, upon his return, be able to answer the letters asking for preprints.

In Zurich, Saari concentrated all his efforts on Littlewood's conjecture and finally he got it. But this time he told no one. He wanted to be absolutely sure. For this he had to understand several more details. Back in Evanston, he went through every line again. He would always remember that decisive, turbulent day, but not on account of the tornado.

Between 1971 and 1973 Donald Saari published three articles, in which he essentially proved that *in the n-body problem, the set of initial data leading to collisions of any type has Lebesgue measure zero*. Are collisions then not worth studying? If it is extremely improbable that collisions occur among all possible orbits, shouldn't mathematicians direct their attention toward other issues? The superficial answer is *yes*, but deeper reflection suggests the opposite. Let us explain the reasons for this.

The importance of collisions is not so much for their own sake, but because they are *singularities*. The structure of phase space changes drastically in the neighborhood of a singularity, so that solutions that do not end in singularities but merely come close to them are likely to behave strangely. We have seen this already in the numerical simulations of the Pythagorean problem illustrated in figure 3.4. In a sense, the collision point corresponds to an infinitely deep well in phase space. As the bodies approach, potential (gravitational) energy is given up to kinetic energy and their relative velocities increase dramatically, becoming greater the closer the approach. The set of these solutions, which glance and swing by singularities, has positive Lebesgue measure, and they cannot be neglected anymore. A proper understanding of singularities helps one understand these orbits as well.

Several years later Saari returned to the idea of that night, at the beginning of his career. He improved his earlier work on Littlewood's conjecture by showing that the set of solutions leading to collisions form lower-dimensional *manifolds*. On one hand this implies that collisions have Lebesgue measure zero, on the other it provides a more geometrical understanding of them. He finally proved that *in the four-body problem, all solutions with singularities have Lebesgue measure zero*. This means noncollision singularities are also rare in this case. It is still unknown how large the set of noncollision singularities is in the *n*-body problem for *n* bigger than four, but many believe it to be negligible too.

Saari still did not know if noncollision singularities could actually occur. This problem recaptured his attention many years later when he started to supervise a new graduate student from China, Zhihong Xia. We will shortly describe this and the fruitful collaboration that followed.

It is interesting that Saari's research interests include another and ostensibly quite different topic, *mathematical economics*. He applies the theory of dynamical systems to economic and political issues. For example, he has proved that the problem of apportionment of congressional seats in the United States House of Representatives gives rise to mathematical chaos. This implies that whatever method is chosen, there will be advantaged and disadvantaged states. It is impossible to find a fair partition.

Donald Saari currently holds the Arthur and Gladys Pancoe Chair of Mathematics at Northwestern University. It is particularly fitting that the donor, Arthur Pancoe, received a Master's degree in 1951 for a thesis on Painlevé transcendents.

REGULARIZING COLLISIONS

It is a considerable honor to be invited to speak at the International Congresses of Mathematicians. As we stated in chapter 2, these congresses take place only once every four years. They comprise all areas of mathematics and are attended by several thousand mathematicians from almost every country in the world. There are few chances to present one's work to such an extensive audience of one's peers. In the summer of 1978, Helsinki hosted the Congress, and Richard McGehee was among those invited to speak. His presentation concerned a new method of analyzing collision singularities in celestial mechanics.

The story of this breakthrough dates from 1970 when McGehee began to work on triple collision solutions of the rectilinear three-body problem, in which, as we noted above, the mass points move along a straight line. He felt that one must understand simple cases before attacking more complex situations. Easier problems offer an excellent test bed to develop new methods and ideas. The rectilinear three-body problem, although the simplest among n-body problems with n greater than two, is still hard enough that it remains incompletely understood even today.

In 1969 McGehee had been fortunate enough to be offered a postdoctoral position at the Courant Institute of New York University. Though born and raised in San Diego, he had grown used to the unpleasant, humid climate of the Midwest during his four years as a graduate student in Madison, Wisconsin. The prospect of spending twelve months in New York did not bother him at all. He was eager to join the famous research group at Courant. It was indeed the right environment for the twenty-six-year-old Californian, and this period at the Institute would have a decisive role in his future career. His first important encounter was with the mathematics promoted by Jürgen Moser, the co-creator of KAM theory—to which chapter 5 is devoted. That year Moser lectured on the Sitnikov problem, which we discussed in chapter 2. He described Alekseev's results on chaotic oscillations. Shortly after this, McGehee attended a presentation given by one of Moser's graduate students, Howard Jacobowitz, who described a paper on the *regularization* of triple collisions in the three-body problem, published almost thirty years earlier by Carl Ludwig Siegel in the *Annals of Mathematics*.

To *regularize* a collision means to extend the motion beyond collapse through an elastic bounce, without loss or gain of energy. The case of binary collisions had already been solved by a Finn of Swedish origin, Karl Sundman, during the first decade of the twentieth century, and published in a final form in 1912 in *Acta Mathematica*. He showed that such an extension of the motion through the collision is always possible. Siegel asked

Plate 3.5. Richard McGehee. (Courtesy of R. McGehee)

what happens if *three* particles collide. Can the motion be continued after collapse in a meaningful way? Siegel showed that this, in general, is not possible. More precisely, he proved that for almost all values of the masses an *analytic* solution that goes through the triple collision cannot be found. (*Analyticity* is a technical property that relates to how the functions describing the solution are defined: the details need not concern us here.)

McGehee decided to work on the Sitnikov problem (fig. 2.11). Within a couple of years he had obtained several important results. To do this, he invented a clever transformation that turns the differential equation describing the particle C into a more revealing form. These new variables "desingularize" the solution as it escapes to infinity. In his 1973 book, *Stable and Random Motions in Dynamical Systems*, Moser uses McGehee's transformations in his account of some of these results.

McGehee's real interest remained in the regularization problem. He was intrigued by Siegel's result, but did not see any connection between it and the real dynamics of planetary motion. "Collisions of celestial bodies are not elastic," he thought. "Why bother what happens after collision?" He wondered how the problem could be posed in order to have a real physical meaning.

Celestial Billiards

Charles Conley had been McGehee's Ph.D. thesis supervisor at the University of Wisconsin. Along with Stephen Smale and Jürgen Moser, Conley was one of the most influential figures in dynamical systems theory in the West during the 1960s and 1970s. He was unusually generous and free with his ideas, thereby launching a number of mathematical careers. Many of his students went on to make important contributions. In several cases he encouraged others to publish work which had its roots in his own ideas.

In 1970 Conley became interested himself in the subject of triple collisions. He and McGehee kept in touch, and during the latter's postdoctoral year in New York they resumed working together. After several discussions they formulated a precise question: *Is the motion of the particles at collision continuous with respect to the initial data?*

We have remarked that scientific research begins when something is unclear. But one cannot find an answer without a question. Often half the battle is in formulating the right problem. This question was indeed fundamental. It also made perfect sense in the context of planetary motion. In physical terms, it asked, Is a collision trajectory the limiting case among orbits that come close to, but avoid, collision? For binary encounters this was known to be true. The particles in the two-body problem can collide *only* if the motion is rectilinear. In all other cases, as Johann Bernoulli had shown, the bodies travel on elliptical, parabolic, or hyperbolic orbits. The elastic rebound can be seen as a natural limit of a sequence of ellipses becoming flatter and flatter, degenerating into a line, as shown in figure 3.8.

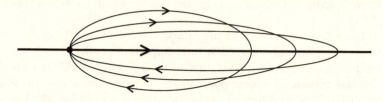

Figure 3.8. The rectilinear orbit is a limiting case of elliptic motion.

Plate 3.6. Charles Conley. (Courtesy of Catharine A. Conley)

The rigorous mathematical proof of this fact was given by another student of Conley, Robert Easton, in 1971. Easton's method employs *mathematical surgery*, a technique from topology, and is based on transformations used by the Italian mathematician Tullio Levi-Civita in the second decade of this century.

This work showed that the physically appropriate question is not whether the motion can be (analytically) continued beyond the collision, but if it can be extended so that neighboring orbits look similar to it. It is helpful now to return to the example of the river and think of the flow around a rock standing in midstream. All the flow lines part smoothly to pass on either side, except the single central one, which leads directly to the rock and a "collision." Thus, in celestial mechanics, Conley and McGehee asked if the singular, triple-collision orbits fit into their environment in phase space, so that they mesh nicely with orbits coming close to such collisions.

Easton had dealt with binary collisions. What about the triple approach? Conley had a brilliant insight. He suggested comparing the three-body

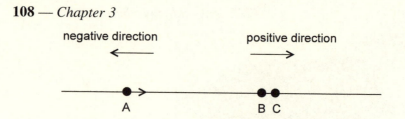

Figure 3.9. Conley's arrangement of the billiard balls, which suggested that the triple collision is not regularizable with respect to neighboring orbits.

problem to a game of billiards. Any experienced pool player knows that if three balls collide simultaneously, then it is difficult to predict their behavior after collision. But this is the result of experience rather than a proven fact that provides a clear idea of what happens. Conley wanted to turn this intuition into a precise mathematical statement. He imagined arranging three balls along a line, as shown in figure 3.9, and proposed the following pair of thought experiments.

Ball *A* is first hit in the positive direction toward the stationary balls *B* and *C*. *A* collides with *B* and returns in the negative direction, while *B* moves to the right and collides with *C*. Following this second collision, *B* also returns, following the path of *A*, while ball *C* moves in the positive direction away from *A* and *B*. There is a qualitative change in configuration: the balls *B* and *C* start out close to each other, but after two binary collisions, *A* and *B* are close, while *C* goes on alone. An exchange of pairs has occurred.

He then imagined exchanging balls *B* and *C*, so that *C* now lies between *B* and *A*. Since *B* and *C* were close together before and are still close, the initial data has barely changed. But, repeating the game with the new initial positions, now *C* leaves the initial pair with *B* and forms a new pair with *A*, while *B* escapes.

McGehee immediately understood the implication. The final motions of *B* and *C* were completely different in the two situations, in spite of only a small change in the initial conditions. (For smaller changes one must place *B* and *C* closer together, so that the two binary collisions approach a triple collision.) This was clear proof of sensitive dependence of the solutions with respect to the initial data. Conley also remarked that this might imply nonregularizability of the triple collision, since two solutions coming close to triple collisions—starting at almost the same point in phase space— behave so differently after a triple approach.

The rectilinear pool game was suggestive, but now McGehee and Conley had to solve the same problem with gravitational forces between the balls, instead of simple elastic impacts. They attempted at first to answer the question of regularizability for the rectilinear three-body problem. McGehee was supposed to prove that if the particles come close to the triple collision, then the central one leaves one pair and forms another. Conley would take care of the rest. At that time they did not know that the *lemma*, or step in the proof that McGehee had taken responsibility for, was not only the key to their projected paper but also a property of utmost importance for future development of the theory of singularities in celestial mechanics.

In the fall of 1970 McGehee accepted a tenure-track position with the School of Mathematics at the University of Minnesota in Minneapolis. He continued his work on the Sitnikov problem and started to think about the lemma concerning the regularization of triple collisions. Although he made nice progress on the first subject, his attempts to solve the second showed only how difficult it really was. He tried several ideas, hoping to get some insight into the problem, even going back to the transformations used by Tullio Levi-Civita over fifty years before in the regularization of binary collisions. All his attempts failed. A full year later he had still not found a solution.

A more impatient person than Charles Conley might well have become frustrated. At this point, he had completed his part of the work and was waiting for the proof of the lemma. He had even drafted a paper, leaving room for McGehee's contribution. The whole argument was based on this missing result: nothing could proceed without it. Now McGehee felt under pressure from two directions. On the one hand, he had to publish: no university grants tenure without clear evidence of good progress in research. On the other, he felt uncomfortable and embarrassed at having to make his former adviser wait. In spite of continuing efforts, he was still lost in the maze of the problem.

They say luck is with the strong. One day he remembered a trick used in some of his previous work. In 1965, as a graduate student, he had become interested in finding a *cross section* for the Poincaré map in a restricted three-body problem (cf. fig. 1.11). He managed to determine and analyze it, and decided to send the result to a journal. But the hope of seeing his maiden publication in print was brief. While writing the paper he found a reference to an article published in the twenties by a student of Birkhoff. McGehee's year of work had been anticipated, for the earlier article had already settled the issue. Now, six years later, he realized that, if combined with the method used for the Sitnikov problem, his old idea might work for the regularization question. He found some new transformations that generalized his previous ones. And this method indeed proved to be the key needed to unlock the solution. It is known today as *McGehee's transformation*.

When Conley read the proof in manuscript and understood the ideas, he realized that his own contribution to the problem was insignificant.

Always generous with his colleagues and students, he called McGehee to ask him to publish the work himself. The paper, with McGehee as sole author, appeared in the European journal *Inventiones Mathematicae* in 1974.

McGehee's idea was to introduce a new coordinate system that "blows up" the triple collision singularity, allowing one to cut out the singular point and paste a smooth manifold in its place. Let us be more precise. The triple collision corresponds to a point in phase space, like the rock in the river's course, where an orbit stops. Blowing up this point is like looking at it under a microscope with infinite magnifying power. What McGehee saw was a distorted sphere with four horns pulled out to infinity and the points at their tips deleted. We sketch this in figure 3.10. He named this surface a *collision manifold*.

McGehee went on to study the flow on the collision manifold. Though it has no direct physical significance, knowledge of it implies understand-

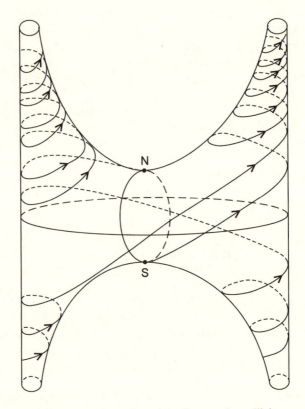

Figure 3.10. One possible configuration of the flow on the collision manifold of the rectilinear three-body problem.

ing of how solutions behave *near* the triple collision as they swirl around the horned sphere. This is a consequence of the continuity of the solutions with respect to initial data. Orbits coming close to this strange sphere must behave similarly to the orbits on it, at least for a time, just as those on tight ellipses follow closely the elastic rebound in the two-body problem. The flow winds upward around the collision manifold and has only two fixed points, which we can call the north pole, N, and the south pole, S. A triple collision occurs when an orbit in phase space "arrives" at S, and a triple ejection occurs when an orbit "leaves" N.

Moreover, McGehee's transformations slow down time and velocities near the singularity so that an orbit takes an infinite amount of (fictitious) time to reach the south pole S. The instant of collision is replaced by an infinitely long interval and the (fictitious) velocity decreases to zero at the moment of impact. These distorted variables provide a magnified and more detailed picture of what happens near the triple collision. This strange manifold is extremely useful in the study of collision and near-collision orbits; it helps one understand the limiting behavior of colliding particles as well as the structure of orbits coming close to a triple collision and then glancing off.

Many new properties have been discovered with the help of this striking geometrical insight. One of them concerns how a particle coming close to a triple collision escapes with high velocity, as the numerical results obtained in the sixties had predicted. In his paper, McGehee used collision manifolds to show that, for certain special values of the masses, the triple collision is not regularizable. In a subsequent paper, he states that this actually holds for all masses except a set of measure zero. This resolved the question that he and Conley had set out to study.

The most important feature of this method is its wide range of applicability. Using it, many other well-known properties have found fresh, more elegant proofs. It extends to all collision questions of the n-body problem and to many other investigations on singularities in the theory of differential equations. It is hard to imagine studying singularities in celestial mechanics today without using McGehee's transformations.

As we have remarked, Charles Conley was responsible for launching a number of mathematical careers, and he and his students developed many important geometrical and algebraic methods in dynamical systems; McGehee's work is but one example; "isolated invariant sets" and "Conley indices" are others, which are now being used in both conventional and computer-assisted proofs of subtle properties of differential equations. However, when once asked about the most natural language for describing the structure of phase portraits, Conley is reputed to have said something like: "For me, the most natural language for discussing the structure of phase portraits is still English."

ENCOUNTERS AT A CONFERENCE

The next step toward settling Painlevé's conjecture was taken by Richard McGehee and John Mather, from Princeton University. They made extensive use of the new collision manifold method. Their collaboration occurred in a rather unusual fashion. Both mathematicians were interested in the problem, but they had no contact until the summer of 1974 when the Battelle Conference, held in a small, picturesque place near Seattle, brought them together. (That they had not met prior to this was not especially unusual; what happened next was.)

Although McGehee had tried to demonstrate the existence of a noncollision singularity while working on the triple-collision problem, he was unable to complete the proof. Mather had learned about the question from a completely different source: a discussion with his colleague Ed Nelson in 1965. In the 1930s in Princeton, John von Neumann was laying the foundations of *operator theory*, and the problem had arisen as a special property of a certain kind of function. To understand the function, one needed to know the dimension of the set of singularities; this was in turn related to whether noncollision singularities do indeed exist. Though differently motivated, Mather and McGehee had a common goal.

Charles Conley introduced Mather and McGehee to one another at Battelle. They began right away to discuss their mutual interests, hoping to collaborate at some time in the future. But already that evening, each realized that he had answers to the other's questions. The next day they continued the discussion, and the day after. It was like a puzzle in which each player has the pieces the other is missing. In less than a week they put the structure together. On the last day of the meeting, John Mather showed McGehee a written draft of a joint paper, which they later published in the proceedings of the Battelle Conference.

What was the result they found that week? They proved the existence of an orbit that becomes unbounded in finite time in the rectilinear *four-body problem*. This did not solve Painlevé's conjecture, since their noncollision singularity is obtained only after passing through binary collisions. (It was already clear from one of Saari's results, described above, that the first singularity to occur in the rectilinear problem must be a collision.) By the process of *regularization*, they extended the solution beyond these collisions. From the physical point of view this means the motion continues after an elastic bounce, without loss or gain of energy. After infinitely many such binary collisions, a noncollision singularity is reached.

The process is pictured in figure 3.11. The masses m_1, m_2, m_3, m_4 are suitably chosen, and the initial positions and velocities are taken such that m_1 and m_2 stay close together, forming a binary system. The particle m_3 oscillates between the particle m_4 and this binary. The collisions of m_2 with

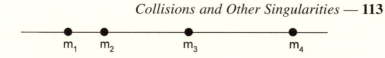

Figure 3.11. The example of Mather and McGehee.

m_3 and of m_3 with m_4 are regularized through elastic bounces that take place at successive instants t_1, t_2, ..., t_k, ... After every triple-collision approach of m_1, m_2, and m_3, the particle m_3 is expelled with significantly higher velocity toward m_4. The collision with m_4 returns m_3 again toward the binary, and so on.

The sequence of collision instants converges to a finite value t^* as k goes to infinity. As this happens, the binary tends to minus infinity (i.e., to the left "end" of the infinite line) and m_4 goes to plus infinity (i.e., to the right "end" of the line), while m_3 continues to travel back and forth between the two. In the meantime, the distance separating m_1 and m_2 decreases, while the potential energy lost by the binary is converted to kinetic energy of the particle m_3. In this way the velocity of m_3 can grow sufficiently fast for it to make the trips back and forth in decreasing time intervals. The construction and verification of this apparently simple scenario required an infusion of new ideas and rather sophisticated techniques.

In fact Painlevé himself had suggested that a noncollision singularity might occur in a situation similar to the one described by Mather and McGehee. Painlevé had not specifically considered the rectilinear case, but motions close to a line. In his case, binary collisions could have been avoided. But how?

In 1978 Robert Sheldon, a student of Howard Jacobowitz, published a paper in which he proved that *if* a solution of the four-body problem does have a noncollision singularity obtained without encountering binary collisions, then the configuration must tend asymptotically to a rectilinear one. But in spite of several attempts to prove that such a solution exists, we still do not have an example to show that Painlevé's conjecture is true in the four-body problem.

From Four to Five Bodies

A famous open question is like a rich mineral lode: everyone wants to exploit it. It was thus with Painlevé's conjecture. After the Mather-McGehee example, the collision manifold technique seemed to be the key needed to unlock the solution. To obtain the desired singularity, one would evidently have to use several triple or multiple collision approaches. But Sheldon's work and the ensuing failure to find noncollision singularities in the four-body problem suggested that this would require at least five bodies.

Increasing the number of particles implies an increase in the number of variables and the dimension of phase space, and consequently much more difficult computations. Choosing to work with the planar or the spatial problem instead of the rectilinear one implies an even larger increase in the number of variables. Any attempt to imitate the proof of the rectilinear four-body case led to the same trouble in dealing with binary collisions. The next step, consistent with keeping the complications at a tractable level, would be the planar five-body problem.

Joseph Gerver, whom we left some pages back listening to Carl Simon's lectures in Berkeley, had joined the faculty at Rutgers University in 1983. Before that, he had spent some time teaching in Hawaii where he had plenty of time to think about mathematics and enjoy fishing. He often re-called his years in New York when, as a graduate student at Columbia University, he had solved an old conjecture due to Riemann. It had been quite unexpected. One of the final homework assignments in a course he was taking was really tricky. None of his fellow students could solve it, but after several days and nights Gerver found a proof. His supervisor and the other members of the department were astonished.

Those heady days seemed far away, for now Gerver was working on a much more difficult question: Painlevé's conjecture. He had thought from time to time about this problem after having learned of it in 1970, in Carl Simon's course at Berkeley. Realizing Gerver's interest in the problem, Simon had written to Donald Saari about this enthusiastic young student, and an exchange of letters between Saari and Gerver began. One day, visiting the Chicago area, Gerver dropped in at Don's office on the Northwestern campus. They spent a fruitful Saturday talking about the *n*-body problem.

By 1984 Gerver seemed close to solving the conjecture, and in order to establish priority, he submitted a paper for publication. In this article he produced an heuristic example in the planar five-body problem, which might give rise to a singularity without any collision taking place first. He didn't have a publishable proof. His argument comprised over a hundred pages of rough computations. It is hard to say if they were correct. Almost no one is prepared to spend weeks to check whether interminable sheets of letters and figures, which appear to contain no new methods or ideas, provide a correct proof or not. Gerver himself was not pleased with his justification, so he preferred to exclude the computations and restrict his presentation to some original qualitative arguments supporting his main idea.

This was Gerver's scenario: Consider five particles in the plane having masses $m_1, \ldots, m_5$, with $m_3 = m_4$, m_2 greater than but of the same order of magnitude as m_3, m_1 much smaller than m_2, and m_5 much smaller than m_1. Take initial positions as shown in figure 3.12. At first m_1 is in a roughly

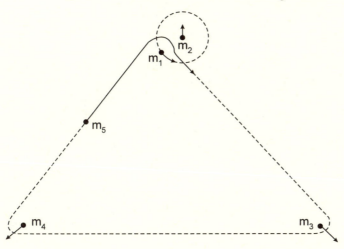

Figure 3.12. Gerver's heuristic example of a noncollision singularity in the planar five-body problem.

circular orbit around m_2, whereas m_3 and m_4 are much farther away. The particles m_2, m_3, and m_4 lie approximately at the vertices of an obtuse triangle, which is slowly expanding while maintaining its shape. In the meantime, m_5 moves rapidly around the triangle, coming close to each of the other four bodies in turn, its velocity being much greater than that of m_1.

Each time m_5 passes close to m_2, it picks up a small amount of kinetic energy. This causes m_1 to fall into a lower orbit around m_2 such that the mean kinetic energy of m_1, in its orbit, increases by about the same factor as that of m_5. A small fraction of the kinetic energy of m_5 is transfered to m_2, m_3, and m_4, causing further expansion of the triangle, but the increase in speed of m_5 is such that the time required for each circuit of the triangle decreases, in spite of the longer path. The sequence $t_1, t_2, \ldots, t_k, \ldots$, of time instants demarking round trips, converges to a finite value t^*, at which m_5 will have traveled an infinite number of times around the triangle. In the meantime, the triangle has become infinitely large. The mechanism creating this rapid expansion is due, as one might by now expect, to the repeated triple collision approaches of m_1, m_2, and m_5.

Unfortunately, Gerver was unable to prove that this plausible scenario actually occurs in the five-body problem. It is not clear that m_5 can be made to pass on each circuit at the right place through the right angle, just by assigning appropriate initial conditions. Something else had to be done in order to reduce the huge amount of computation. But at that time, he could not see what it was.

THE END OF A CENTURY'S QUEST

Zhihong Xia had received his Bachelor's degree in astronomy from Nanjing University at only nineteen years of age and was hoping to continue his studies in the United States. His interest in celestial mechanics had begun a couple of years before, and he had eagerly read all the relevant papers he could find on the *n*-body problem. He particularly admired Donald Saari's work, and so wrote to ask if he could become a graduate student at Northwestern.

After the partial opening of China to the West in the 1980s, American scientists whose work was known in that country would sometimes receive several requests each month from prospective graduate students. It was impossible to act on or even personally answer all of these. Chance had it that Xia would first begin his studies elsewhere in North America, but his desire to work with Saari in Evanston persisted and he decided to try again. This time he was lucky. Saari is a kind, warm person, full of understanding and concern for those who need help. He promised to see what he could do. After a few months Xia moved to Northwestern University, and Donald Saari was to become his supervisor and friend.

The route from China to North America and finally to the campus in Evanston was not easy for a young man like Zhihong Xia, but after some time he adapted to his new surroundings. He asked people to call him Jeff, since, for most Americans, his given name was hard to pronounce and memorize. Xia aimed to return to China after receiving his Ph.D. He knew how difficult it would be to obtain an academic job in North America. U.S. universities often assume that the language barrier makes foreign-born Ph.D.s poor teachers and Chinese students find it particularly hard to gain such posts.

So Xia set to work. He was bright and willing to devote twelve or fourteen hours a day to research, seven days a week. Moreover, he now had a wonderful mathematical goal: to settle Painlevé's conjecture. He had of course been introduced to it by Saari, who realized that Xia had the potential to attack such a difficult question. His initial idea was to find a suitable solution to a six-body problem. Although this at first appeared promising, it failed in the end because of a certain transversal intersection of orbits on the collision manifold. There was clear evidence, from numerical work, that the solution he was seeking would eventually manifest a noncollision singularity. But the analytical procedure didn't work, and Saari would have never approved a thesis with a numerical argument as "proof."

That summer Xia recognized that the features he wanted to preserve from the six-body problem would remain unchanged in a certain type of

Plate 3.7. Jeff Xia. (Courtesy of Z. Xia)

five-body problem. He immediately sketched the proof, but the details were tedious and would take years to complete.

Some Ph.D. thesis advisers give their students problems and leave them to work alone; others prefer to collaborate closely, feeling that this is the best way to pass on their knowledge and experience. The collaborative style was especially productive in this case. Xia was courageous, in his naiveté trying to scale the most dangerous heights. Saari, with a lifetime of mathematical experience, tried to encourage restraint and prevent him from making mistakes. Several other related questions came up and were answered in the course of Xia's work. Some of them were solved with Saari's

help. They wrote a fine paper together on a certain type of asymptotic behavior in the rectilinear *n*-body problem. Throughout, Xia continued to
work on the noncollision question.

During this period, Harry Pollard came to speak at Northwestern University. Saari knew that one of his former supervisor's dreams was to see a
solution of Painlevé's conjecture. At the party he threw in Pollard's honor,
Saari introduced him to Xia. Pollard was pleasantly surprised to hear that
the young man was quite close to obtaining a proof, but unfortunately Pollard would not live to read the paper. A few weeks after this visit, his health
deteriorated. Within months he had a stroke and, soon afterwards, he died.

Deeply moved by Pollard's death, Xia went back to work with even
greater passion. The dream of solving the problem was not his alone. It had
been a goal for several generations of mathematicians, a pyramid built over
ninety years, and he, Zhihong Xia, was close to placing the final stone.
What feeling could be more rewarding?

In 1987 Xia anounced that he had completed the proof of Painlevé's
conjecture. He gave a talk in Princeton on the subject, and Charles Fefferman, a Fields medalist, urged him to submit the paper to the premier journal, the *Annals of Mathematics*. Xia wrote a preprint of almost a hundred
pages, and sent it to several experts. At the same time he submitted the
paper for publication, following Fefferman's advice.

There was nothing left to do but wait. Xia knew that his expository style
was awkward. It was difficult to express his ideas, he disliked writing
things down, and English seemed the hardest language on earth. He failed
to justify statements that appeared to him intuitively obvious. He was too
inexperienced to realize that such a style drives referees crazy. Had he
known better, he would have spent more time polishing his presentation
and thus saved months of uncertainty and tension for himself, and work for
the referees.

After two years he received an ambiguous answer. The referees could
not decide whether the proof was correct or not. In any case, the problem
was too important to approve publication as the paper was currently written. At least one crucial statement lacked a convincing proof, several others
suffered from clumsy, unclear formulations. Under no circumstances could
the paper be published in this form in a premier journal.

This sort of difficulty is by no means unusual in research. Deeply involved in technical details, it is almost impossible to put oneself in the
place of a reader and provide a clear and faultless presentation from the
beginning. This is another reason why the refereeing system is used to
improve the quality of scientific publications.

In the intervening years Xia left Evanston. He had originally intended to
return to teach in his hometown of Nanjing but, on being offered the posi-

tion of Benjamin Pierce Lecturer and Assistant Professor of Mathematics at Harvard, he decided to accept and remain in the United States. He continued to try to close the gaps and improve his presentation. Some people, finding it hard to believe that a student could have solved such a celebrated problem, began to doubt the proof, but Xia remained confident. He submitted a revised version of the article to John Mather, who was an editor of the *Annals*. Mather resolved to take the decision to publish or not on himself, but how could he find the time to read such a difficult piece of work with the requisite care? At this point it occurred to him to organize a seminar at Princeton, in which he would present each week a part of Xia's proof. What could be more instructive for himself and his students than such a seminar? He knew it would be worthwhile even if the proof was wrong. Sometimes one learns more from mistakes.

The seminar ran through the fall of 1991. It was a fine opportunity for some of the graduate students at Princeton to enter the laboratory of mathematical creation, in a complicated but beautiful subject. At the end of the term Mather delivered his verdict: the proof was correct. The letter of acceptance was perhaps the finest present Jeff Xia had received. The paper appeared in the summer of 1992 in the *Annals of Mathematics*.

When the notification of acceptance finally came, Xia was no longer at Harvard. He had moved to the Georgia Institute of Technology in 1991. He found that he preferred Atlanta. The Dynamical Systems Institute provided an excellent working environment for someone who was barely twenty-nine. His wife could continue her studies without having to pay the high tuition fees at Harvard. This was a good place for both of them, and Xia knew he could tackle other hard problems. But he missed Evanston, and in the fall of 1994 he returned there to a tenured professorship at Northwestern. In a later chapter we shall describe some of the work he did in Atlanta. Now let us see how Jeff Xia resolved Painlevé's conjecture.

Perhaps not surprisingly, he considered a *spatial* five-body problem. In spite of the increased number of variables, he simplified the analysis by fixing certain symmetries. He took two pairs of bodies, all having equal masses $m_1, \ldots, m_4$, plus a fifth particle, m_5, of small mass in comparison with the others (see fig. 3.13). The paired bodies move in highly eccentric orbits on two parallel planes, the lower pair rotating clockwise and the upper counterclockwise, with equal and opposite angular velocities. The motion of the fifth, small particle is restricted to a line perpendicular to these planes and through the common mass center, so the total angular momentum of the system about this line is zero.

The behavior of the small particle is essential to obtaining the noncollision singularity, for it oscillates between the two binaries, producing

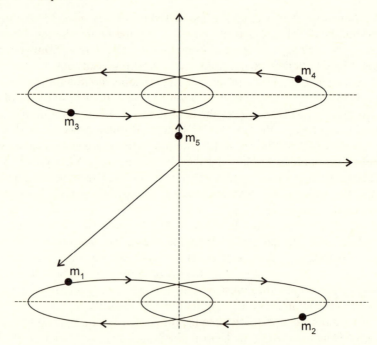

Figure 3.13. Xia's example of a noncollision singularity.

an unbounded motion in finite time. We will describe this part of the scenario in detail. Suppose that the particle m_5, coming from below, crosses the line connecting m_3 with m_4 just before these particles reach their nearest approach: m_3, m_4, and m_5 are then close to a triple collision. The body m_5 goes a little above the line joining m_3 and m_4, while the particles m_3 and m_4 are at their closest approach. Then m_5 is strongly attracted backwards. It crosses the line $m_3 m_4$ again as these point masses start to separate. This separation reduces the retaining force on the small particle, which consequently moves rapidly downward toward the other binary system. At the same time the action-reaction effect forces the binary $m_3 m_4$ to move upward relative to the plane of the lower binary. The encounter is now repeated, in mirror image, for the lower binary $m_1 m_2$. Iterating this procedure with closer approaches and correspondingly higher accelerations for m_5, the two binaries are forced to escape to infinity in finite time.

Simple though this scenario sounds, it is difficult to carry through for Newtonian gravitational laws. Since the motion is to become unbounded in finite time, the acceleration of the small particle must grow infinitely large. To produce the requisite slingshot effect, the point masses in each binary must approach closer and closer, making it hard to guarantee that binary collisions do not occur. Here the structure of stable and unstable manifolds of the collision points is used in a subtle way to eliminate "bad" sets of solutions at each stage. Those which lead to collisions on the next encounter or for which m_5 arrives at the wrong time are removed from the set of candidate solutions. In addition to "aiming" m_5 correctly, the angular velocities of the binaries have to be reduced a little at each step, so that the encounters can become close enough. All this requires a most delicate touch. After repeating the process for infinitely many encounters, one is left with a Cantor set of initial conditions that lead to solutions behaving as we have described.

How can one remove infinitely many unsuitable sets and still be left with something? Xia had to show that, at each step, a finite patch of solutions survives. The reader will recall that we used a similar, but simpler, construction in chapter 2 while building the Cantor set of points which never escape from Smale's horseshoe (fig. 2.6). We also described the construction of the middle-third Cantor set earlier in this chapter (fig. 3.7). Xia used a similar construction in this much more complicated case. In fact, Cantor sets are rife in this subject: in the collinear example of Mather and McGehee described earlier, there is also a Cantor set of initial conditions with the desired behavior.

In this way Jeff Xia showed that there *are* initial data such that collisions can be avoided and the binaries driven to infinity in finite time. He revealed a rare capacity to overcome, on the one hand, technical difficulties, and to manipulate, on the other, a vast quantity of mathematics. He introduced a number of new ideas, but above all, his proof is a wonderful example of collective scientific effort. He brought to bear everything he knew about the *n*-body problem, using a host of results and techniques developed by others before him. It took almost a century to assemble the edifice that supports this lovely conclusion. Xia's theorem is at the apex, but, without the stones laid by his predecessors, there would be no building.

In the summer of 1993, a joint meeting of the Canadian and American Mathematical Societies was held at the University of British Columbia in Vancouver. The occasion was used to present the first Blumenthal Award, to be made every four years in recognition of distinguished achievements in mathematics. The selection committee included Vaughan Jones from Berkeley (himself a Fields medalist in 1990), Robert Langlands from the Institute for Advanced Study in Princeton and the Centre de Recherches Mathématiques in Montréal, Gregori Margulis of Yale University, Peter May from the University of Chicago, and Wilfried Schmid of Harvard. Zhihong Xia was honored as the first recipient for his proof of Painlevé's conjecture.

A Symmetric Digression

One of the problems Xia considered in the course of his existence proof for noncollision singularities was the case of *isosceles* motion. An important part of his analysis was based on what happens when the particle m_5 comes close to one of the binary pairs. In Xia's solution the subsystems of particles m_3, m_4, m_5 and m_1, m_2, m_3 describe isosceles triangles throughout the duration of the solution. By this we mean that the shapes of the triangles with the three masses at their vertices is such that two sides always have equal length. The size can change over time but equality of length remains, for m_5 lies on the vertical line and the binaries orbit symmetrically about it, so that the horizontal components of the forces just balance. Fortunately, much of the analysis of the isosceles three-body problem had already been done by other researchers, and Xia could draw on it.

There are two isosceles three-body problems, one in space, which Xia needed, and a more restricted planar one in which all three particles move in a fixed plane. The planar problem had been studied intensively since the end of the last century and, as we described earlier, in 1961 Sitnikov considered the spatial isosceles problem, seeking an example of oscillatory solutions. He proved that if two particles of equal masses move in elliptic orbits in a plane and a third particle, of negligible mass, moves on the symmetry axis perpendicular to the plane of the other two, as in figure 2.11, then one can find initial data such that the small particle oscillates up and down, its motion becoming unbounded. The oscillations become larger and larger and ultimately unbounded. This was the first example of an oscillatory motion in any *n*-body problem, but—in contrast to the solutions Xia was seeking—it took infinite time to become unbounded.

Jürgen Moser, whom we met earlier at the Courant Institute, considered Sitnikov's problem, and the extensions to it proposed by Alekseev, in a wider context. He was not only seeking oscillatory solutions, but also wanted to describe all possible behaviors of the small particle. In 1973 he published the book entitled *Stable and Random Motions in Dynamical Systems*. In it he showed the connection between possible solutions of Sitnikov's problem and the shift map used by Smale, as we described in chapter 2. To any shift map he attaches a solution, and conversely. Roughly speaking, this means that the small particle on the symmetry axis can in fact move in any manner we specify. We must only choose suitable initial conditions. As we have seen, this can be done with the help of symbolic dynamics.

The most recent results in the isosceles three-body problem are primarily concerned with triple-collision and near-triple-collision solutions. Robert

Devaney of Boston University treated the planar case in 1980 and 1982, while Richard Moeckel (another student of Conley) of the University of Minnesota has worked on the spatial case since 1984. Both have used McGehee transformations and a qualitative analysis of the flow on and near the collision manifold. Their papers suggest that the motion of the general three-body problem near the triple collision is chaotic in the sense described in chapter 2. This shows how difficult it is to choose precisely the right initial data such that any particular desired motion takes place.

Roger Broucke of the University of Texas performed some numerical investigations of the isosceles three-body problem in 1979. Irigoyen, Lacomba, and Losco in 1980, and Simó and Martinez in 1988 obtained further analytical results. But in spite of all these articles, some of them revealing remarkable and interesting properties, the isosceles problem is far from completely understood today.

An Idea at Dinner

It had been a day full of interesting presentations and, as often happens at conferences, the discussions continued long after the last talk. The 1985 meeting of the American Mathematical Society in Anaheim, California, had brought together researchers in all areas of mathematics. Collaboration is often initiated at such events, people from different fields exchange ideas, talk about issues of common interest, and go home better informed and wiser than they came.

Joseph Gerver went for dinner that evening with several other people. He knew some of them, others were new acquaintances. The restaurant was pleasant and quiet, with candlelight, good food, and wine. Various things were discussed: mutual friends, academic politics, and inevitably, mathematics. The usual question is, "What are you working on?" Everyone was telling his own story, his challenges, difficulties, and achievements. Gerver started to talk about the Painlevé conjecture. He soon captured general attention, for it was easy to describe the problem in unsophisticated terms. He explained his attempts to squeeze the proof out of the five-body problem and outlined his reasons for believing this should work.

Everybody was impressed. It is not common to learn about a ninety-year-old problem that is so simple to describe and yet so hard to solve. Some of those around the table began to think about Gerver's story, while others went on to talk of something else. Presently Scott Brown asked, "Why don't you try to construct a solution having a sort of rotational symmetry?" There was a moment of silence, and then Gerver's eyes lit up. "Yes," he replied, "that's a great idea! I will certainly try it."

On the way back to the hotel Gerver figured out how to use the symmetry

Brown had suggested in order to improve his five-body example. That night he immediately started to work on the computations. It was not easy to see whether it would lead to something significant, but he had hopes. Eager to get home and to compare his notes with earlier calculations, Gerver felt closer than ever to the solution of the problem that had obsessed him for the last fifteen years.

Over the following months he tried and tried and tried again, but something was always missing. Several times it seemed he might have got it. He would see a ray of hope on the horizon, but always a mistake would cloud it over. Finally Gerver realized there was no way he could create a noncollision example within the framework of the five-body problem. It was a bitter moment. He felt exhausted. The problem had begun to obsess him, so he decided to put it aside and take a vacation.

But he could not shrug it off quite so easily. One morning he awoke with a new idea. Why should he stick to the five-body problem? Rotational symmetry applies perfectly well to any number of particles. Why not increase the number and give himself more freedom? It would require additional computations, but it just might work. He started once again, devoting enormous amounts of time and energy to it. This time he did not give up. The struggle with the problem and with himself had resumed.

Then in 1987 the news came. Zhihong Xia was to speak on his proof of Painlevé's conjecture at Princeton, and Gerver saw an announcement of the lecture. Who was this person? Gerver had never heard of him and at first did not want to believe it. He found it hard to accept the fact that someone else might have beaten him. After some reflection he thought, "It is only an announcement. Who knows if the proof is correct?" He decided to call Don Saari, who would surely know about it. As soon as he learned that Xia was Saari's student, Gerver knew the race was over. But, like a brave soldier cut off from the main force, he continued to fight, trying to convince himself that nothing had changed. And finally he succeeded: a few months afterwards, he found another, different proof.

In January 1989 Gerver submitted his paper for publication to the *Journal of Differential Equations*. It was accepted in November of the same year, and appeared in January 1991. In spite of publishing the article before Xia's, in his introduction Gerver forthrightly recognizes Xia's priority in giving a first example of a noncollision singularity in the *n*-body problem. Nevertheless, Gerver's work provides the first approach to the planar case.

In celestial mechanics, the bare fact that a particular kind of solution exists is perhaps not as interesting as the details of its construction, and so it is well worth describing Gerver's example. He took $3N$ particles in the plane, with initial positions as in figure 3.14. The number N has to be chosen sufficiently large in order to make the example work. $2N$ of the

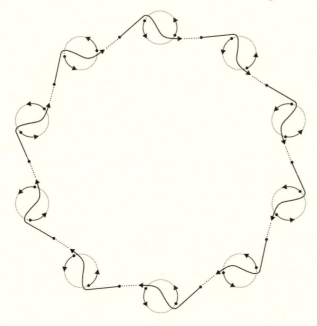

Figure 3.14. Gerver's example with rotational symmetry in the plane.

particles are arranged in N pairs in almost circular orbits, and all have equal masses. The center of mass of each binary lies at one of the vertices of a regular polygon. The other N bodies have small, equal masses and move rapidly from one pair to the other, as the arrows in figure 3.14 indicate. When each small particle comes close to a binary, it takes kinetic energy from the pair and transfers some momentum to it, forcing the binary to move into a tighter orbit and at the same time increase its distance from the center of the polygon.

Continuing this process for a suitably large number of particles, and choosing appropriate values of masses and initial velocities, the size of the configuration can be shown to increase after each close encounter of a small particle with a binary. The sequence of intervals between each encounter and the next converges to a finite value, while the diameter of the system becomes unbounded in finite time. This implies the existence of the noncollision singularity.

Before ending this story we should mention a remarkable insight into the problem that came from someone who never published a word in the celestial mechanics literature: Robert F. Williams, a topologist and dynamicist, now at the University of Texas in Austin. Long before Xia and Gerver, he suggested how one might construct solutions with noncollision singularities, but he never pursued the idea. It happened at a conference mentioned earlier, which was held in 1971 at the University of Bahia in Brazil's oldest city, Salvador, a place of beautiful churches, squares, and gardens. At that time Williams was on the faculty of Northwestern University. During the coffee break after his colleague Don Saari's talk on noncollision singularities, he suggested to Richard McGehee and other people nearby that in a type of solution like the one to be described twenty years later by Jeff Xia, a noncollision singularity might occur. McGehee was reluctant to accept this proposal and convinced him that the idea would not go far. After a few minutes Williams came up with another example, very similar to that ultimately given by Gerver, but again it did not receive approval. At that time everyone was inclined to believe that noncollision singularities did not exist. The property of rapid separation after a triple approach had not yet been discovered. Since his proposal had been greeted with little enthusiasm, Williams did not think about the problem again, although he subsequently became widely known for his work on strange attractors and knotted periodic orbits in chaotic dynamical systems of another sort.

It is difficult to compare the two solutions to Painlevé's conjecture obtained by Xia and Gerver. Each is interesting and valuable in its own way. Xia pioneered a new direction, using sophisticated mathematical apparatus and bringing fresh ideas into the field of celestial mechanics. Xia's paper has been read with great care and there is no doubt today about its correctness. In his work, Joseph Gerver managed to overcome the difficulties of this subtle problem using the classical tools of nineteenth-century mathematics. His account may be less appealing to read because of the immense volume of computations, but Gerver must be admired for his tenacity in performing them. Both people achieved remarkable feats in an old and difficult field, where significant new results are hard to come by.

Even today Painlevé's conjecture is not completely exhausted. Xia's construction can be generalized to n particles, for any n bigger than five, yet there is no hope to adapt it to the four-body problem. Whether noncollision singularities exist in that case or not is still an open question. On the other hand, the attempts to prove the conjecture have enriched classical mechanics and mathematics with new methods and ideas that are already being used in other applications. As with most good questions, Painlevé's conjecture has left behind a small world of intellectual accomplishments, and we can see its legacy as a lovely piece of scientific culture created through the combined efforts of several generations.

4.

Stability

Is the solar system stable? Properly speaking, the answer is still unknown, and yet this question has led to very deep results which probably are more important than the answer to the original question.

—Jürgen Moser

I̲T SEEMED a typical evening. After dinner, Jean le Rond was in the habit of reading his correspondence and, if time allowed, replying to some of the many requests he received. He was sitting at his desk, writing, when a servant entered the room carrying a calling card and a letter on a silver tray.

"Oh no, not again," came his master's protest after unsealing the letter. "Another reference for one of those 'young talents' recommended by members of high society." He turned to the man. "Please explain that I cannot receive him."

Jean le Rond d'Alembert was a retiring man as well as a great mathematician. He found flattery unpleasant and would not put up with anyone's company solely on account of wealth or social rank. As the illegitimate son of an aristocrat, he had learned bitterly that the nobility were not necessarily rich in spirit. Shortly after his birth, his mother had abandoned him on the steps of the little Chapel of Jean le Rond in Paris. His foster parents baptized him Jean le Rond, and he later added d'Alembert to his name.

The letters of reference arriving that day were written on behalf of an eighteen-year-old from Beaumont-en-Auge, in Normandy. In them, a number of influential people praised the mathematical genius of this young person, requesting that d'Alembert receive and help him. They could have known little about the character of the famous mathematician, or they would not have sent such recommendations. Jean le Rond was not impressed by the opinions of the ill-informed. An event that occurred later in his life reflects his personality rather well.

Frederick the Great offered d'Alembert the most prestigious position a scientist could have imagined at the time: the presidency of the Berlin Academy. Anyone but d'Alembert would have been honored to accept. In his letter of refusal, Jean le Rond related how, like most other people, he had originally craved honors and wealth, but had ended by giving them all up. In exchange, he had received perfect health and the peace of mind, which, he believed, are the only rewards a philosopher should desire.

The incident of the talented young man soon faded from d'Alembert's mind. A few days later, however, he received a long and impressive essay on the general principles of mechanics, signed with a name he seemed to have seen before: Pierre Simon. Unable immediately to place it, he realized after a few seconds that this had been the youth with references from influential people. "This person needs no recommendations!" he thought, smiling. "He knows perfectly well how to introduce himself." Without delay d'Alembert sent an answer, proposing an appointment for the following day.

A Longing for Order

There is a desire for eternity that every human being expresses at one time or another. No matter how futile we may believe this to be, we nonetheless strive to prolong our lives as much as possible. The first requirement for this is safety, the assurance that our world is not dangerous. We hope that our cities and towns will be safe and our countries not involved in war. We want our children to grow up in peace and happiness. More fundamentally, we trust that our physical environment will not suffer radical change.

Ever since humankind has learned that its home is a small, vulnerable planet in the greater solar system, scientists have raised the question of its continued existence. We depend entirely on the energy from the sun; the distance to it proves to have been ideal for the creation and evolution of life. Had it been a little smaller or greater, life as we know it on the blue planet would probably have never arisen.

Yet long ago, astronomers observed that the motions of the earth and of the other planets are not perfectly regular and periodic. This prompted them to ask if the earth will always maintain its orbit around the sun. Might it leave its present neighborhood and drift farther out into the solar system, or perhaps even worse, collide with some other large body such as an asteroid or comet? Today we have ample evidence that collisions with objects of modest size have occurred in the past, causing changes in climate and weather, severe enough, perhaps, to wipe out the dinosaurs. In the summer of 1994 the fragments of comet Shoemaker-Levy crashed into Jupiter. Astronomers have recently discovered the comet Swift-Tuttle in the neighborhood of that planet; preliminary calculations based on estimates of its orbital elements indicate a 1:40 probability of collision with the earth in the twenty-second century. What else can we expect in the future?

Centuries before such specific evidence came to light, questions like these had led mathematicians to the notion of *stability*, a fundamental concept in the study of nature. In principle, stability refers to perturbation phenomena. If slight changes occur in a dynamical system, will it maintain its state or not? There are, however, dozens of definitions of stability, each having a different meaning, depending on the context. Many of them are related; some imply or are implied by others. At the end of chapter 2 we mentioned Poincaré's recurrence theorem, which represents a weak type of stability in which we ask only that the system repeat its behavior from time to time. We shall return to it in a few pages. In figure 1.6 of chapter 1 we sketched phase portraits of stable and unstable equilibria, and described a stricter and more direct notion of stability in the context of a pendulum hanging down or standing up in an inverted position. Now we shall consider what might be the simplest kind of stability to grasp as a physical notion, but perhaps the hardest to deal with mathematically: the stability of the planets in their orbits around the sun.

The solar system is said to be stable if there are no collisions among the bodies in it and if no planet can ever escape from it. In terms of the n-body problem, this means there are no singularities of any type and no particle ever leaves the system, even in the infinite future or past. (As soon as we move from the physical solar system to the mathematical universe of the n-body problem, we forget that the earth is only some four or five billion years old. Orbiting point masses can persist forever in the ideal world of differential equations.) This is rather a weak definition of stability. It requires only boundedness and permanence. There must be no catastrophes, but nothing is said about how close the planets should remain to their present orbits. (Nor does the n-body problem pay attention to the fact that the sun will eventually exhaust its nuclear fuel and swell into a red giant, incinerating the earth in the process.)

The only way to tackle such a question with mathematical precision is to write down and study the equations describing the Newtonian model of the solar system: a ten-body problem having one large mass—the sun—and nine small ones. We find that planets rotate around the sun on *almost* elliptical orbits. Why are they not exactly elliptical? The explanation within Newton's model is that the orbit of each planet is slightly influenced by the motion of the other planets, so that it does not describe a perfect ellipse, but only remains close to one. An n-body problem of this type is called, naturally enough, a *planetary problem*.

It is therefore clear that no rigorous study of stability of the solar system could have come before Newton. In fact, it was almost a century after the publication of *Principia* that the first pertinent results appeared. Nevertheless, the general question of stability, irrespective of the solar system, had arisen long before that, in antiquity.

The Greek natural philosophers Aristotle and Archimedes approached the problem from two different viewpoints. The former looked at it kinematically. Taking the balance beam and objects lying on planes as examples, he imagined the kinds of forces that might disturb an equilibrium state and the motions that would ensue. In contrast, Archimedes espoused a geometric view. His studies of floating bodies demonstrated the importance of the shapes of objects and the positions of their centers of gravity. In retrospect, one might say that Aristotle took the experimental viewpoint (although he may never have done practical experiments), while Archimedes constructed a mathematical model. At that time the very notion of stability was still lacking. It was apparently the Latin poet Lucretius who first used the term "stabilitas" in his poem "De Rerum Natura" (On Nature), in a passage where he remarks on the stability of light and heavy objects.

The Greeks and Romans lived in a physical world dominated by friction. It was commonly thought that a force had to be applied to keep a body moving. Not until Galileo's "Dialogue Concerning Two New Sciences" (1638) was the idealization of smooth, friction-free bodies introduced. Galileo was also probably the first "physicist" to do careful experiments. His work on balls rolling down inclined planes contributed to the formulation of Newton's laws of motion. At around the same time, the concepts of equilibrium and stability were clarified and separated. Torricelli, the inventor of the barometer and a younger contemporary of Galileo, was apparently the first to clearly describe *unstable* equilibria. Thus far the focus was on static equilibria. Subsequently, Huygens, who was interested in pendulum clocks, introduced dynamical ideas, but all these early investigators evidently thought of stability with respect to *finite* disturbances. Infinitesimal perturbations, so crucial to modern studies of stability, had to await the invention of calculus in the late seventeenth century.

We have already seen in chapter 1 that Poincaré's discovery of chaos arose from his attempts to prove its exact opposite: stability. As frequently happens in mathematics and science in general, new things often appear while one is looking for something else. The notion of stability was not only important in itself, but also acted as a catalyst in the progress of science. In this sense it can be said to have inspired and produced the revolution of chaos.

Now it is time to tell the story of earlier attempts to prove the stability of *n*-body problems.

THE MARQUIS AND THE EMPEROR

Pierre Simon was only twenty-four years old when he submitted his first memoir to the French Academy of Sciences. His paper contained a remarkable result about the planetary problem, for in it he claimed that the solar system is stable.

Born in 1749 to a peasant family in Beaumont-en-Auge, Department of Calvados, in Normandy, Pierre Simon quickly distinguished himself as a gifted student of mathematics. He attended courses as an external student at the Military Academy in Beaumont, and it seems that he also taught mathematics there for a while. It was during this period that Pierre met several rich people who took him under their protection, appreciating his rare ability for the exact sciences. At the age of eighteen he was already a professor at l'École Militaire in Paris, as a result of the recommendation and influence of d'Alembert. Pierre Simon was subsequently appointed to a chair at the renowned École Normale Superieure. He lived through the Revolution and the reign of Napoleon, and after the restoration of the monarchy he became Marquis de Laplace under Louis XVIII. His prestige in eighteenth-century French science was to be equaled only by that of Lagrange.

Unlike d'Alembert, Laplace had a complex and difficult personality. On the one hand, he strove throughout his life to hide his humble origin, but on the other, he helped a number of young, impoverished men who were attempting to enter mathematics. He was strong and persistent in research, but weak in public life, inclined to accept titles and honors even when he did not merit them, and not always willing to acknowledge his debt to others, including Lagrange, whose work we shall shortly come to.

By 1785 Laplace was already a full member of the French Academy. A particular event that year would come to change his life by diverting it toward politics. In the course of his normal academic duties he had to examine a sixteen-year-old candidate who impressed him from the outset. This teenager's name was Napoleon Bonaparte. Later, after Napoleon's rise to power and in some part because of it, Laplace became rich and influential. But although he moved in the highest circles of society, he continued working on mathematics all his life with the same success and enthusiasm of his youth.

Pierre Simon's interests were broad. He did fundamental work on probability theory, summarized in his great *Théorie analytique des probabilités*, and, incidentally, wrote a delightful essay on the laws of chance. For our purposes, his masterpiece is the five-volume treatise entitled *Traité de mécanique céleste*. Over twenty-six years in the making, it reviewed all

Plate 4.1. Pierre Simon Laplace. (Courtesy of Gauthier-Villars, Paris)

the important achievements in the field since Newton, and contained many new ideas and results. It was in fact Laplace who introduced the term "celestial mechanics." For those unable to follow the details and abstractions of this vast book, his subsequent *Exposition du système du monde* proved one of the most successful popularizations ever to appear.

The Emperor Bonaparte was himself a competent mathematician; some results in elementary geometry bear his name today. Shortly after the

Traité appeared, he asked Laplace to visit him to discuss the ideas in it. The main goal of the book was to explain, through Newtonian gravitation, the existence and the phenomena of the entire solar system. Laplace was an atheist, so he never mentioned the name of God, as had been done in almost all of the earlier papers and books treating the subject, starting with Galileo and Kepler. Napoleon was a great soldier and accustomed to winning. Powerful and feared by most, he thought himself above all others. This time he wanted to best Laplace, so he pressed the attack: "You have written this huge book on the system of the world, without once mentioning the Author of the Universe!"

Laplace may have been equivocal in public life, but when the issue was science, he always spoke plainly. He never denied what he stood for. His book was based on a single axiom: particles of matter attract each other following the gravitational law. With this hypothesis he could explain everything. Nevertheless, to be completely frank under such circumstances might have caused him serious trouble. Though recognizing the danger, he gave a straightforward and honest answer: "Sire, I had no need of *that* hypothesis."

Napoleon looked deeply into Laplace's eyes. The old scholar returned the emperor's look with dignity and pride. Nothing in the world would have deflected Pierre Simon at that moment. After an awkward silence, Bonaparte changed the subject. (Subsequently, when told by the emperor of this exchange, Lagrange is reported to have remarked: "But it is a beautiful hypothesis; it explains many things.")

Laplace's first major contribution came in his memoir on stability of 1773. The central theorem in that paper contains the following statement: *In the first-power series approximation of the eccentricities, the major axes of the planets have no secular terms.* What does this technical language mean? The *major axis* of a planet is defined as the segment *AB* in figure 4.1. It is the longest straight line connecting two points of the ellipse that forms the planet's orbit. As we have pointed out, in the presence of more than two bodies, the length of this segment generally varies with time. The orbit is an inconstant ellipse whose size can vary. If the variation is too large, the planet's trajectory might become unbounded.

While the major axis measures the size of the ellipse, the *eccentricity* characterizes its shape; it describes how close it is to, or how far from, a circle. Large values of eccentricity correspond to long, flat ellipses. To take the first-power series approximation of the eccentricity means to provide a rough estimate of it. A second-power series approximation generally gives a better estimate than the first approximation, but worse than the third approximation, and so on. Such a technique is called *perturbation theory*, for one takes a known solution to a simpler but similar problem and sucessively modifies it to better approximate the true solution, which one cannot

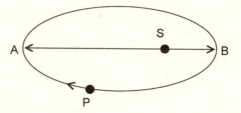

Figure 4.1. The major axis *AB* of a planet *P* is the major axis of the ellipse supposed to be its orbit around the sun *S*.

calculate exactly. To apply perturbation methods, one needs a small parameter in terms of which to make the series expansion. Here the eccentricity—the slight deviation from circular orbits—was the natural choice. We
recall that much of Poincaré's work, described in chapter 1, had to do with
series approximations and their convergence.

The *secular terms* denote the linear (or sometimes faster) growth of the
time variable in certain expressions. Since we are concerned with behavior
as time goes to infinity, it may well happen that expressions containing
such terms do not remain bounded. If *no* secular terms appear in an *exact*
calculation, rather than in a power series approximation, we may conclude
stability, for the quantities characterizing the (perturbed) ellipse all remain
constant. If secular terms do appear, either stability or instability may
ensue. More precisely, if these terms somehow conspire and cancel, so that
the final expressions remain bounded, then we have stability. Otherwise the
system is unstable. In either case, computation of only approximate estimates cannot settle the question conclusively.

The status of Laplace's result now becomes clear. In showing that no
secular terms appear in a first approximation, he had not so much proved
stability as failed to demonstrate instability in solutions to the mathematical model. Moreover, even if stability could be proven, it would refer to the
n-body problem and not directly to the solar system itself.

It was nonetheless a pioneering result in celestial mechanics. Until this
time, the scientific community had not expressed a clear opinion on stability. Newton, Euler, and others had encountered serious problems in their
attempts to understand the orbit of the moon. Believing that the motion of
the planets posed much more difficult problems, they had not attempted the
larger question of stability itself. Laplace's detailed calculations and his
introduction of the seeds of a general method made this possible. His development of perturbation methods therefore has great historical importance:
he blazed a path that many have followed since. Perhaps the most dramatic
application was the discovery of Neptune in the mid-nineteenth century,
following the predictions of Adams and Le Verrier.

Music of the Spheres

The next contribution came sooner than expected. Joseph Louis, Comte de Lagrange, entered the picture in 1774. Born in Turin in 1736, of both Italian and French origin, and Laplace's senior by thirteen years, his story is surprisingly similar to the latter's. Joseph Louis had to begin work early, as his father had lost his wealth following some foolish financial speculation. At the age of sixteen Lagrange was appointed geometry professor at the Turin Artillery School. Within four years he had sent Leonhard Euler a paper on *isoperimetrical properties* (the property of two or more geometrical figures having boundaries of the same length). In it, he introduced ideas that he would go on to develop into the foundations of the *calculus of variations*. Today this has grown into an independent field of mathematics. Lagrange went on to become director of the scientific section of the Royal Academy in Berlin, and was ultimately elected to the Institut de France, the successor of the French Academy of Sciences.

Lagrange's life and career were as successful as those of the younger Laplace. But above all, Joseph Louis always had a good word for everyone, friend or stranger. Without regard to his own situation, he helped many people in need. He did not complain about Laplace's failure to credit his contributions to the former's work. He was interested in much beyond mathematics, having the erudition of a reflective thinker who tries to understand the world as a whole. He disliked no one, and his heart was as large as his knowledge.

Early in his career, Joseph Louis became attracted to the problems of celestial mechanics. He was an analyst, the first such to introduce analytical language to the study of mechanics, as opposed to the geometric methods inherited from Greeks, which Newton had used in *Principia*. Lagrange worked at various times on the three-body problem, some equilibrium solutions of which bear his name, as we shall see at the end of this chapter. He devoted many years to this and other problems in celestial mechanics.

In 1764, at only twenty-eight years of age, he won the Grand Prix of the Paris Academy of Sciences for a memoir on the libration of the moon. Observed from earth, the moon seems to show always the same face; her dark side was seen only after the first space probes had photographed it. In fact, slight rotations to the left and to the right cause her to reveal a little more than half: about 60% of her surface in all. These oscillations are called the *libration*. Lagrange was able to explain this phenomenon using Newton's gravitational law.

He won a second Academy prize in 1766 for solving a yet more difficult problem. He explained certain inequalities in the six-body problem defined

Plate 4.2. Joseph Louis Lagrange. (Courtesy of Gauthier-Villars, Paris)

by the sun, Jupiter, and Jupiter's four largest satellites: Io, Europa, Gany-mede, and Callisto, the only ones known at that time. In 1772 Lagrange received another prize from the Paris Academy, for a memoir on the three-body problem. The groundwork was in place for his attack on the stability question. He embarked on it in 1774.

When struggling with a difficult problem, Lagrange tried to dedicate himself to it continually. This was particularly hard, since mundane events and duties, unavoidable in his position, would constantly interfere with the solitude he required. At one of the grand receptions given at the time by

Frederick the Great, Joseph Louis was, as usual, absentminded. For many weeks now his thoughts had focused on the stability problem, but he had been unable to overcome some technical difficulties. In the glittering, candle-lit rooms, numerous people greeted him and wanted to talk; polite as he was, he did not rebuff anyone. He would have liked to concentrate on his thoughts; unfortunately this was clearly the wrong place for it. Finally, a chamberlain announced that the concert was about to begin. Lagrange smiled. He had been waiting for this moment.

A friend saw the expression of contentment come over his face and asked: "Do you really like music *that* much?" Joseph Louis couldn't refrain from laughing. "Yes, I do," he answered. "I like it because it isolates me. I hear the first three measures; at the fourth I distinguish nothing; I give myself up to my thoughts; nothing interrupts me; and it is thus that I solved more than one difficult problem."

That night was no exception for Lagrange. He plunged into the silence of the music, free to explore the inner world of his mind, undisturbed in his private process of discovery. He recalled the unresolved details of the problem and went deeper and deeper, coming closer and closer to an answer. At the end of the concert he had made an important step forward. He was eager to return home to make some final computations.

Between 1774 and 1776 Lagrange extended Laplace's result on stability in the following sense. He proved that *for all order approximations of the eccentricities of the ellipses (given as orbits of the planets), for all order approximations of the sine of the angle of the mutual inclinations, and for perturbations of the first-order with respect to the masses, the solar system is stable in the sense that secular terms do not occur.* This statement is even more technical and complicated than Laplace's, and we shall not attempt to explain it in full. The essence is that approximations to the eccentricities of *all* orders were included along with other more accurate estimates, and that everything was done with respect to a first-order approximation obtained by perturbing the masses. Lagrange's series approximations employed *three* small parameters: eccentricity, the inclinations of the planes of the orbits, and the ratio of the planetary masses to that of the much larger sun.

This new approach generalized the result of Laplace, making use of better approximations at a finer level than those employed by the Marquis. It did not completely solve the problem: the estimates offered an improved, but still an imperfect model of physical reality. However, it was a remarkable result, and it suggested that the solar system might indeed be stable.

Lagrange remained interested in celestial mechanics throughout his life. This latest success encouraged him to study the motion of the moon and that of cometary perturbations. He wrote further memoirs for which he won other prizes. In addition, he continued to develop analytical methods in mechanics in more general terms, and in 1788 he published his masterpiece, *Mécanique analytique*. He had been determined to carry this out since the age of nineteen. Here he introduces for the first time the idea that mechanics is the geometry of four dimensions: three spatial coordinates and a temporal one. This original viewpoint has of course become famous since Einstein founded his theory of relativity upon it. Lagrange also developed the perturbation methods that Laplace had used in his stability calculations, and made them generally accessible. Indeed, his description of the ideas is clearer than that of many modern textbooks. The German physicist Ernst Mach described *Mécanique analytique* as "a stupendous contribution to the economy of our thought," and William Rowan Hamilton, whom we shall meet in the next chapter, called it "a scientific poem." Nonetheless, in view of our general theme, it is ironic that Lagrange took pride in his reduction of mechanics to a part of pure analysis, and that his great book contained not a single figure.

Briefly, the perturbation method goes as follows. Suppose we have a differential equation which can be completely solved, giving explicit formulae for the solutions as functions of time (see chap. 1), and we wish to solve another equation which is "close" to this one in the sense that its vector field is a small modification of the original. Naturally, we expect its solutions to be likewise close to those we have already found. Each such solution will contain *parameters*, equal in number to the dimension of state space, and determined by the initial conditions of the particular solution in question. For the original problem, all of these remain constant; essentially, they tell us the correct mixture of fundamental solutions necessary to match the given initial data.

The perturbation method of Lagrange allows these parameters to vary with time, the key assumption being that they do so *slowly*, because the difference between the new problem and the original one is small. (Thus the method is sometimes paradoxically called *variation of constants!*) This often allows us to express the solutions as the product of rapid oscillations and a slow drift, and one can derive equations for the slowly varying terms by averaging the accumulated perturbations due to the rapid oscillations. These averaged equations are usually simpler than the original ones, thus they can be solved and their solutions can be used to obtain a better approximation of the "true" behavior than the original unperturbed solution provides. In the celestial context, the rapid oscillations might be the annual planetary rotations about the sun and the drift, the precession (slow rotation) of an elliptical orbit which would remain fixed in a two-body universe.

Like Laplace, Lagrange's scientific skills were also greatly appreciated by Napoleon Bonaparte. The emperor considered Joseph Louis to be "the lofty pyramid of the mathematical sciences." Lagrange was made a senator, a Count of the Empire, and finally a Grand Officer of the Legion of Honor. For his many great achievements he gained full recognition not only in science but also within society at large. Sadly, his private life was not always happy. His first wife died young, and Joseph Louis did not seek to remarry. The following two decades in his life were marked by long depressions, probably brought on by uninterrupted periods of work. And then a miracle occurred. The teenage daughter of the astronomer Lemonnier was so touched by Lagrange's unhappiness that she asked him to marry her. Joseph Louis, fifty-six at that time, understood the gravity of such a step and the sacrifice the young woman would be making, but finally he agreed. Contrary to all expectations, the marriage was a wonderful success. It transformed the remaining twenty years of Lagrange's life.

Although shaken by the excesses of the terror that followed the Revolution, Lagrange was active in attempts to improve university education in France. In 1794 he was a founder of what would become the famous École Polytechnique. He also taught at the École Normale, where Laplace was his assistant.

In the spring of 1813 he became ill. Realizing that the end was near, he called his closest friends to bid them farewell. He calmly explained how he had observed the gradual diminution of his physical and intellectual strength, how he felt himself slipping away without pain or regrets. "Death is the only absolute repose of the body," he said. "I wish to die; yes, I wish to die and I feel a pleasure in it, but my wife did not want it. At these moments I should have preferred a wife less good, less eager to revive my strength, who would have let me end gently. I have had my career; I have gained some celebrity in mathematics. I never hated anyone, I have done nothing bad, and it would be well to end. . . ."

Two days later, on the morning of 10 April 1813, Joseph Louis Lagrange closed his eyes forever. He was mourned by his many friends, by France, and by the world of science. His body lies in the Panthéon, along with those of other great statesmen and military heroes. But the universe of his ideas is still alive.

ETERNAL RETURN

Investigations on stability of the solar system did not end with Lagrange. In 1809 another French mathematician, Siméon Denis Poisson, went one step further. He proved that *the major axes of the planets have no purely secular terms in the perturbations of the second order with respect to the masses.* Poisson offered a slightly better approximation than Lagrange and adduced more support for stability, but he too did not completely resolve the problem.

Siméon Denis Poisson belonged to a new generation of French mathematicians who were to usher in the great achievements in analysis of the

Plate 4.3. Siméon Denis Poisson. (Courtesy of Springer-Verlag Archives, Berlin)

nineteenth century. When the stability result appeared, he was twenty-eight years old and poised to become a leader in the theory of differential equations. Like his predecessors Laplace and Lagrange, his taste for mathematical research was combined with an equal talent in physics. He made contributions to mathematical physics, hydrodynamics, electrostatics, magnetism, ballistics, and elasticity. A famous partial differential equation, which bears his name today, is used in the study of stellar dynamics. Changing the sign of a certain constant in this equation, one obtains a model used in plasma physics. (This observation is due to the Russian mathematician A. K. Vlasov, who was to be one of Kolmogorov's teachers, as we shall see in chapter 5.) The same equation also appears in elasticity theory and fluid mechanics, providing a nice illustration of the unifying power of mathematics.

Poisson also proposed a new definition of stability, which differed from that of Laplace and Lagrange. His predecessors had required the major axes of the ellipses to remain within certain limits. Denis Poisson had a different idea. He defined the motion of a system of particles to be stable if its configuration returns close to the initial position again and again. Although equilibrium points such as those in figures 1.6b and d are clearly also stable in this sense, it is a much broader definition and includes other, less obviously stable motions. The recurrence theorem of Poincaré that we mentioned at the end of the second chapter had as its starting point Poisson's definition of stability.

Before leaving French mathematics and moving our story to eastern Europe, let us describe Poisson stability. As we mentioned in chapter 1, the first part of the last volume of *Les méthodes nouvelles* dealt with this notion. One of the major results that Poincaré describes in this section is his *recurrence theorem*. As we saw in chapter 1, this had already appeared in his *Acta* prize memoir. For the three-body problem, it goes as follows. Suppose three gravitating particles have certain initial positions at a given moment. Assume that no collisions occur, and all motions remain bounded. Then, after a sufficient interval of time, the particles will return extremely close to, if not exactly to, their initial positions. But having once returned, we may apply the same argument to conclude that the particles must return *infinitely* often to positions close to these initial ones. Poincaré showed that this property holds for all but a negligible set of initial conditions, i.e., with the exception of a set of zero Lebesgue measure.

This important result, used also in the kinetic theory of gases, subsequently gave birth to the branch of mathematics called *ergodic theory*, to which George David Birkhoff was to make crucial contributions. As we mentioned earlier, Poincaré dedicated almost half of his third volume to this subject. In spite of this, the idea behind the proof of the recurrence theorem is quite easy to understand, so we will briefly describe it.

Consider a container full of water, as shown in figure 4.2, and assume the liquid to be in continuous motion. Let V_0 denote a volume of water at some initial instant t_0. At some later moment t_1, the molecules previously in V_0 will have moved to a new region, V_1. Now, V_1 may have a very different shape from V_0, but since water is an incompressible fluid, the volumes of the two regions must be equal. At subsequent instants of time, $t_2, t_3, \ldots, t_n$, $\ldots$, the molecules will occupy other positions $V_2, V_3, \ldots, V_n, \ldots$. We claim that at some (finite) instant t_m, the corresponding volume V_m occupied by the water molecules must overlap or *intersect* the original region V_0. If this failed to occur, then all the volumes $V_0, V_1, V_2, V_3, \ldots, V_n, \ldots$ must be separate (*disjoint*, in mathematical language), and so the total volume traced out by our moving blob of fluid would keep growing and eventually exceed the volume of the container. (This is a proof by contradiction—a common trick in mathematics.)

In the reasoning above, we did not use the actual measure of the initial volume V_0 at all, so we can take it as large or as small as we wish. Now, if V_0 is very small, then the fact that V_m and V_0 have common points means

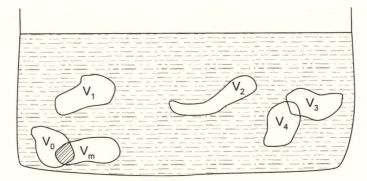

Figure 4.2. Proof of the recurrence theorem.

that some molecules must return close to their initial positions, although it may take very long for them to do so. This is precisely the conclusion stated in the recurrence theorem. In our "proof," we used a three-dimensional container of water to represent phase space, but the argument applies in just the same way to the n-dimensional phase space of any dynamical system, provided the n-dimensional analogue of volume *is* preserved by the flow of the differential equations or under iteration of the map. As we have remarked, phase space volume *is* typically preserved for celestial and mechanical systems, due to the conservation laws of Newtonian mechanics.

It is interesting to note that Poincaré actually used his recurrence theorem in the revised *Acta* prize memoir to deduce the existence of infinitely many homoclinic orbits from one. The theorem appeared in both versions of his paper.

There is one important observation to make. We assumed that the volume of V_0 is positive. If it has Lebesgue measure zero, then our reasoning fails, for the volume traced out by a moving blob of zero size does not grow. In fact, idealizing the water molecules as points, there may exist some that never return to a neighborhood of their initial position. A correct statement of the recurrence theorem must therefore exclude a set of initial data of Lebesgue measure zero. Of course, this does not detract from the value of the result, for the set of cases to be excluded is of negligible size.

In modern terminology we say that the recurrence property is *generic*, meaning that it is satisfied by almost all solutions. This was another novel viewpoint that we owe to Poincaré. Up to his time it was generally supposed that a *single* counterexample disproves the rule. Poincaré disagreed. He was not bothered by even *infinitely many* counterexamples as long as they are unlikely to appear. Who cares about them if their measure is negligible in comparison with the immensity of all possibilities, and if all the others follow the rule? A central concern of the theory of dynamical systems today is that of finding such generic properties of mathematical objects and events in phase space.

Poincaré's insight becomes even more remarkable when we realize that the concept of measure had not even been defined at that time. He evidently had an intuition for it and was able to use this in *Les méthodes nouvelles*. According to Immanuel Kant, notion without intuition is empty, while intuition without notion is blind. If we are to believe the German philosopher, Poincaré made his discoveries with closed eyes. But how well he could see!

Perturbing the World

The war continued. After many centuries of Ottoman suzerainty, Romania had declared its independence. It was a bitter winter, but good news about the fighting on the Balkan front, south of the Danube, was reaching the interior of the country. There was little doubt that the Romanian Army was close to a final victory.

The morning of January 31, 1878, seemed at first like any other. People in Bucharest waited in lines to buy the newspaper, hoping to read of a new success in this struggle to protect their young and fragile liberty. But they had a big surprise. The leading article on the front page did not concern the war. Instead the headlines announced a cultural victory: Spiru Haretu had been the first Romanian mathematician to obtain a doctoral degree in mathematics in Paris. The young man had successfully defended his thesis the previous day. Such events were rare at that time; the number of doctoral degrees in all fields was small. It was natural that this, the first of its kind, should receive special attention and be treated as a matter of national pride. Indeed, the following day the newspaper *Românul* compared Haretu's achievement with the Romanian Army's recent victory at Plever.

For over a thousand years, the three principalities that form Romania today had helped to shield central and western Europe against invasions from Asia and the Ottoman empire. Romanians had to struggle continually for the right to live in their own land. At last this situation seemed to be coming to an end.

While Romania shielded one of its gates to the Orient, central and western Europe flourished culturally and intellectually. The first institutions of higher learning were established as early as the eleventh century in Bologna and in the twelfth in Paris and Oxford. In contrast, the earliest Romanian universities were founded only in the nineteenth century. Having encountered such difficulties in their fight to survive, Romanians appreciated culture and considered it a first good of their nation. (An old Romanian proverb says "Ai carte, ai parte," meaning, "If you study, you will be better off.") From this sprang the feeling of pride and the importance attached to the country's first doctorate in mathematics.

Currently, over one thousand mathematics Ph.D.s are awarded each year in the United States and Canada, with probably a similar number in Europe, so we may find it hard to imagine how Haretu's achievement prompted a national outburst of joy. Romanians were happy simply on account of the event. Who knew what subject the thesis treated? Ordinary people needed only to hear the message, not its content. But some who read the newspapers that day realized that Haretu had also shaken the scientific world. He was the first to raise serious doubts regarding the stability of the solar system.

At twenty-six years of age, Spiru Haretu had submitted to the Sorbonne a thesis, "On the Invariability of the Major Axes of the Orbits Described by the Planets." In spite of its dry and unrevealing title, this remarkable piece of work went a step beyond Poisson's memoir. Haretu proved that *in taking the third-power series approximation of the masses, secular terms do occur in the values of the major axes of the planets*, thus suggesting precisely the opposite conclusion to that of Laplace, Lagrange, and Poisson. Haretu's calculations showed that *the major axes of the orbits described by the planets are not necessarily bounded in time*.

As we have hinted, this statement does not immediately imply instability, for it is not clear how large the variation of the axes can be. It does, however, make it clear that the orbits of the planets might change their shapes and sizes as time progresses.

Haretu obtained his result by considering a yet better approximation than the one taken by Poisson. Others had naturally tried this step, but encountered seemingly insurmountable technical difficulties. The only advances up to 1878 were due to Liouville and Puiseux in 1841 and to Tisserand in 1876, who had considerably simplified Poisson's lengthy proof. Tisserand used an idea of Jacobi from a celebrated paper on the *reduction of nodes*, which had been an important step toward understanding the three-body problem.

The Romanian mathematician also employed these tricks to transform the problem into a simpler one. In modern mathematical terminology, he carried out the *reduction of a Hamiltonian system with symmetry*, using constants of motion to lower the number of independent equations that had to be solved. As we suggested in our discussion of manifolds in chapter 1, this allowed him to reduce the dimension of phase space, so that he was able to push the calculations through to the third-order approximation. We shall have more to say about this technique in chapter 5. At this stage, Haretu found that secular terms *do* occur in the formulas describing the major axes of the planets. He succeeded where all his predecessors had failed.

Plate 4.4. Spiru Haretu. (Collection of F. Diacu)

Spiru Haretu was immediately offered a professorship at the University of Grenoble. However, he felt that Romania needed him more, and his patriotism, mingled with the pride of belonging to a newly independent and victorious nation, prompted him to return to Bucharest—"the little Paris," as this beautiful city was known in Europe before the Second World War. He joined the faculty of the University of Bucharest, and his influence was crucial in building up the prestige of that institution. He helped mold the careers of many students and laid the foundations of a strong mathematical school. This was, however, only one aspect of his public life.

Haretu entered politics after returning to his native country and became almost as successful as Painlevé was to be in France, two decades later. A member of the Liberal Party, he was appointed minister of education at an early age. He held this portfolio several times, under different governments, and in this capacity founded the national elementary educational system. He vowed to create a school in every village, to make culture available to all children, to seek out and promote gifted students, and to convince politicians and investors that a good educational system is the premise for a sound economy. Haretu also reformed the secondary and higher educational institutions by adapting the French model to Romanian realities. With his influence and connections, he opened the door to many young scholars who would later receive their doctoral degrees in Paris. Drawing on the favorable political climate between Romania and France at the turn of the century, Haretu contributed to the rich cultural exchange between the two nations.

As a minister he was unusually popular, taking the time to visit schools in remote villages, talking to teachers, and trying to help with ideas and funds. He would go alone, unannounced, in casual dress, as a friend rather than an official, trying to understand the realities in rural education.

This deep involvement in public life did not leave Haretu the time and the peace of mind to repeat his early scientific achievements. Although he continued to do mathematics, his results were no longer as impressive. Nevertheless, in 1910 he published an interesting and original book, *Mécanique sociale*. This work was an attempt to combine his mathematical knowledge with his deep experience of social matters, by applying the methods used in mechanics to mathematical models of societies. From this point of view he can be seen as a pioneer in a branch of sociology.

For a time Haretu's mathematical work was almost forgotten. His detailed perturbation calculations were among the last to be done on the stability question. His inconclusive result suggested that power series approximations would not provide decisive answers, and, as we have seen, shortly afterwards Poincaré pioneered the qualitative approach. Nevertheless, Haretu's contribution was praised by Poincaré in the first volume of *Leçons sur la mécanique céleste*. Moreover, due to its connection with the *small denominators problem*, Haretu's theorem can be considered as a foundation stone for the great edifice of KAM theory, which we shall describe in chapter 5. In 1958 Jean Meffroy, professor of mechanics at Montpellier, reconsidered Haretu's ideas, and using newer methods established not only the existence of the third-order secular perturbations but also found analytical expressions for them.

Spiru Haretu died at the age of sixty-one, in the same year as Henri Poincaré, leaving behind a sound educational system, a model of which other Romanian intellectuals could be proud. To honor his contributions in celestial mechanics, the International Astronomical Union assigned his name to a lunar crater. In downtown Bucharest, in University Square, there is a statue of a man with a parchment in his hands. The statue has survived the bombardments of two world wars and the fighting of two revolutions. On the pedestal one can read the name: *Spiru C. Haretu.*

How Stable is Stable?

The status of the stability problem for the solar system at the end of the nineteenth century was summarized by Poincaré in an article published in 1898. In spite of some historical inaccuracies, the paper is a useful survey for nonspecialists. "People who are interested in advances in celestial mechanics but who can only follow them from afar," writes Poincaré,

> must experience some surprise in seeing how many times the stability of the solar system has been proved. Lagrange established it first, Poisson demonstrated it again, more demonstrations have come since, more will yet come. Were the old proofs insufficient, or is it the new ones that are superfluous? These peoples' surprise will undoubtedly redouble if one tells them that one day, perhaps, a mathematician may show, by rigorous argument, that the solar system is unstable. This really could happen; there would be nothing contradictory about it, and the old proofs would still retain their value. It is just that they are actually only successive approximations; thus they do not pretend to rigorously bound the elements of the orbits between narrow limits from which they can never escape.

After this carefully qualified assessment, Poincaré did not try to deceive the scientific world or himself by overemphasizing the importance of mathematical applications in practice. He well knew that all this work addressed only the idealized *mathematical model* proposed by Newton. How well this model represents reality was a much harder problem. In 1891, in another popular article on the three-body problem, Poincaré had written: "One of the questions with which researchers have been most preoccupied is that of the stability of the solar system. This is, if truth be told, more a mathematical question than a physical one. Even if one were to discover a general and rigorous proof, one could not conclude that the solar system is eternal. It may, in fact, be subject to forces other than those of Newton." In this, as in so much else, he was prophetic: some quarter of a century later, Einstein proposed the general theory of relativity.

The Qualitative Age

Haretu's paper came near the end of an era in mathematics. The two-century dominance of *quantitative methods* was waning. The work of Bruns on first integrals of the *n*-body problem foreshadowed the end. A few years later, Poincaré's prize memoir was to mark the beginning of the age of *qualitative approaches*. It started in differential equations, where it would in time lead to the *geometric theory of dynamical systems*, but it soon began to enter other fields of mathematics. The older methods and ideas seemed to have reached their limits in various branches of science. Their central role in progress and development was in abeyance. The new ideas had arrived at the right time.

Until the final decade of the nineteenth century, the goal was to obtain exact results, to find formulas that would yield precise numerical predictions, to integrate equations and obtain complete solutions. Mathematicians such as Leopold Kronecker believed that God created the world using integer numbers and, consequently, everything could be reduced to them. This was, in a way, an extension of the idea promoted by the Greek philosophical school of Pythagoras, which sought to explain everything in terms of regular polyhedra: the five perfect solids. Indeed, once we have a language, a set of tools, it is very tempting to try to force the world to fit our method. Nature is, however, not as simple as such systems would have us believe, and dynamics is no exception. The set of problems that can be "completely" solved is depressingly small. Physical phenomena are in general nonlinear and some are even chaotic. We now realize that, apart from a negligible class, most differential equations cannot be solved explicitly. But following Poincaré's advances, mathematicians realized that they could still investigate properties of solutions, *even if they could not find the exact solutions themselves*. This is the essence of qualitative methods.

A typical example is that of stability. If one cannot find the solution of a differential equation in terms of an explicit formula or even estimate a reasonable approximation, one can still try to determine if certain kinds of orbits, known only to exist but not described precisely, are stable or not. Questions like this *can* often be answered, and it is the goal of the qualitative theory of dynamical systems to find methods and results that can help solve such problems.

In spite of two hundred years of efforts to settle the stability question for the solar system, the first attempt to create a fully general mathematical theory of stability was not made until the end of the nineteenth century, by a Russian mathematician: Aleksandr Mikhailovich Liapunov. It may seem strange that people would attempt to solve a problem in science without the necessary tools and without even an agreed-upon definition of what had to be done, but discovery in science, as in the rest of life, is often disorganized. Although our textbooks do not present it thus, rigorous mathematical statements are usually developed only after numerous failures and partial successes in formulating and solving special cases. It was not until Liapunov's doctoral dissertation, presented in Kharkov in 1892 and published one year later, that the general notion of stability for a solution of an ordinary differential equation was defined. While a French translation of his huge paper was published in 1907, and its ideas have long been in

Plate 4.5. Aleksandr Mikhailovich Liapunov. (Courtesy of Nauka, Moscow)

common use, it appeared in English only in 1992, in the *International Journal of Control.*

Mathematical aptitude usually makes its appearance at an early age, and most creative mathematicians obtain doctoral degrees at the latest in their mid- or late twenties. Liapunov was an exception. He did not complete his thesis until the age of thirty-five. But immediately after its publication he was recognized as one of the leading mathematicians of his time.

Aleksandr Mikhailovich Liapunov was born on 26 May 1857 in Iaroslavl. In 1880 he graduated from St. Petersburg University, where he was a student of Pafnuti Chebyshev (after whom is named a set of functions— Chebyshev polynomials—widely used today in computer simulations of fluid flows). Upon finishing his doctoral degree, Liapunov was immediately appointed professor at the University of Kharkov, in Ukraine, and ten years later began working as a full member of the St. Petersburg Academy of Sciences. He did not publish widely, but all his results opened up new directions. Liapunov liked privacy and, after he became an acade-

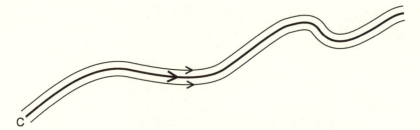

Figure 4.3. Stability in the sense of Liapunov for the orbit *C*.

mician and was exempt from teaching duties, he devoted almost all of his time to research. His rare public appearances outside the university were at the concerts given by his brother Sergei, a composer and collector of folk songs who was for a time director of St. Petersburg's imperial choir. Liapunov's death was as unexpected as it was tragic. In 1918 his wife died of tuberculosis. He could not bear the pain of her loss and shot himself the same day. All attempts to save him failed. He died three days later.

Liapunov's doctoral thesis is the starting point of the modern theory of stability. The definition he gave of this fundamental notion is simple and powerful. Consider a differential equation and a particular solution in phase space, such as the one represented in figure 4.3 by the curve *C*. We say that this solution is *stable* if all other solutions of the same equation starting close to *C* remain close to it for all time. In other words, there is a (narrow) strip in the phase space, containing *C*, such that *every* solution curve having a point in the strip will remain forever inside that strip. This greatly generalizes the notion of stability of a fixed point or equilibrium solution, introduced in chapter 1, for Liapunov's definition applies to *any* kind of solution: steady, periodic, or more exotic ones.

This definition may seem similar to the continuous dependence of a flow with respect to the initial data. But there is also a crucial difference. The continuous dependence property demands only that solutions stay close to each other *for a while*. The closer two solutions start, the longer they will stay together, but this generally holds only for a finite interval of time. Stability requires that *all solutions that are near a stable orbit at some time must stay close to it forever*. This is much stricter than Poisson stability, which only requires occasional returns and does not preclude wild excursions in between.

To say that a solution is *unstable* in the sense of Liapunov means that no strip such as that above exists. Regardless of how close to *C* we start another orbit, it will leave any neighborhood (i.e., any strip) around *C* after a finite time has elapsed. An example is sketched in figure 4.4, in which

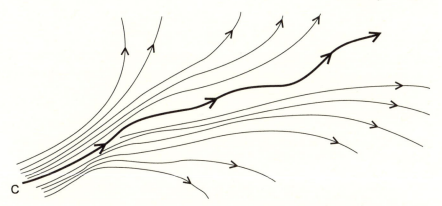

Figure 4.4. Instability in the sense of Liapunov for the orbit *C*.

several orbits leave the strip. To conclude instability, however, it is enough to find *one* such orbit.

Stability is a rare property among solutions of differential equations, and if present, its existence is usually hard to prove. Stability in the sense of Liapunov is also in some sense a "good" property, for it implies the robustness of the physical phenomenon described by the corresponding solution.

The study of stability for a nonequilibrium solution *x* is usually done by looking at a different equation than the one in question, which *x* actually solves. This may seem odd, but it is in fact the natural thing to do. By subtracting $x(t)$ from the quantity defined in the original equation, we obtain a second differential equation for which the time-varying *x* has become an equilibrium. It is much easier to look at what happens in the neighborhood of an equilibrium than in the vicinity of a possibly complicated, time-dependent solution. This simple device shows that the study of stability for an *arbitrary* solution of a given differential equation can be reduced to the study of stability for an *equilibrium* of a related differential equation. Thus, our problem is reduced to the development of stability theory for equilibria.

This trick, which works because we have focused on only a single solution, makes our task much simpler. To see this, let us review the definition of Liapunov stability in the new context. Solutions starting close to the equilibrium point *E* will have to stay close to it for all future time, as figure 4.5 shows: in the new equation, the strip around the orbit *C* has become a little disk or ball around *E*. Now, these perturbed solutions are not generally equilibria themselves, so they tend to wander around *E*, although always remaining near it. In some cases they may get closer to the equilibrium as time passes. When this occurs, we speak of *asymptotic stability*. Asymptotic stability implies (Liapunov) stability, but the converse is not

(a)   (b)

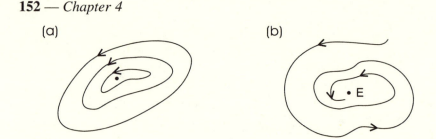

Figure 4.5. (a) Liapunov stability, and (b) asymptotic stability for an equilibrium solution.

true. Consider figures 1.6d and 4.5a, in which orbits circulate nearby for all time, without getting any closer to the fixed point.

The condition of stability—of remaining always in a small neighborhood of the equilibrium—in the definition of asymptotic stability is strictly necessary. In figure 4.6 we show an example of an equilibrium *E* that is *unstable*, even though every solution approaches it in the end. Considering a small neighborhood of *E*, we see that *some* solutions leave this neigborhood; in spite of subsequently returning, they do not stay near the equilibrium throughout, as the definition of Liapunov stability requires. In this case, *E* is not asymptotically stable because it is not even stable. This shows that our definition makes good physical sense: if the quantity in question is the temperature of a chemical reaction, for example, it is little comfort to know that it will eventually cool down if it first exceeds the melting point of the container.

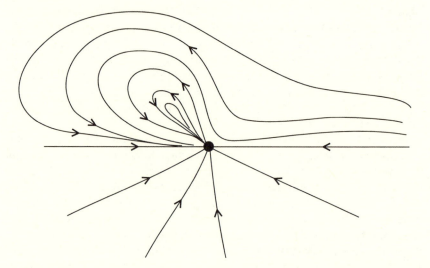

Figure 4.6. An example of an unstable equilibrium to which all solutions tend.

Linearization and Its Limits

It is natural to start the study of stability with equilibria of the simplest kinds of differential equations: linear ones. These are equations defined by linear functions. Linearity is a wonderful and dangerously seductive property in a mathematical model: wonderful, because it is well understood and easy to handle; seductive, because it persuades us to model everything as a linear process, so that we may easily solve the resulting equations.

As we described early in chapter 2, a map or a function is a rule that assigns to each "input" a unique "output." If the input is a real number, the rule might be *multiply by six* (*i.e.*, x goes to $6x$), *divide by four* (x goes to $x/4$), or *take the reciprocal* (x goes to $1/x$). Roughly speaking, a function is *linear* if doubling the input also doubles the output. Thus the first two examples given above are linear, while the last is not. If the linear function has constant coefficients (they do not vary with time), the corresponding differential equation can be completely solved and explicit formulas for the solutions can be given in terms of well-understood functions.

For equilibria of (two-dimensional) linear systems we have three major possibilities:

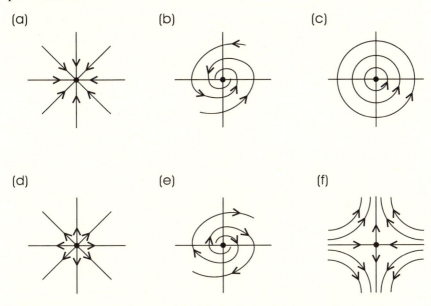

Figure 4.7. All cases of behavior near an equilibrium of a two-dimensional linear system: (a) and (b) the asymptotically stable cases; (c) the neutrally stable case; (d), (e), and (f) the unstable cases.

1. *Asymptotic stability*, when all orbits in a neighborhood of the equilibrium are attracted to it (see figs. 4.7a and b);

2. *Stability*, sometimes called neutral stability to distinguish from (1), when the equilibrium is surrounded by periodic orbits (see fig. 4.7c);

3. *Instability*, when at least one orbit leaves the equilibrium (see figs. 4.7d, e, and f).

With the exception of a set of measure zero, all two-dimensional linear systems belong to one of these types.

This is a fundamental result in the theory of differential equations, but one might easily consider it too special to be taken seriously, arguing that most physical phenomena are described by *nonlinear* equations and so the linear case is insignificant. This would be a mistake. At the end of the sixties and the beginning of the seventies, a Russian mathematician, D. M. Grobman, and an American, Philip Hartman, independently proved that, excepting the case (2) of neutral stability, near any of its equilibria a nonlinear system has the same qualitative behavior as *its corresponding linearization*. What does this mean? We will shortly give a geometric explanation, but first we must explain what is meant by the *linearization* of a nonlinear function, map, or differential equation.

To do this we will consider a function of a single variable, so that we may draw a graph to represent it, as in figure 4.8. Here the "input" is the real number on the horizontal axis, and the "output" is the corresponding number on the vertical axis: the graph provides a geometrical equivalent of a table such as the ones clerks use to compute sales taxes to be added to a purchase price. Sales tax (unlike income tax) is a linear function: one pays

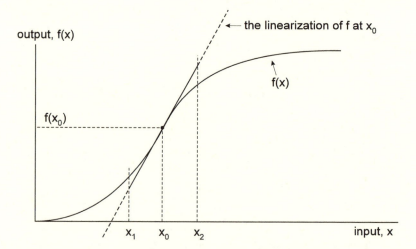

Figure 4.8. A nonlinear function and its linearization at the point x_0.

twice as much tax if one's purchase costs twice as much. In figure 4.8, in contrast, we have sketched a nonlinear function that might represent the response of an electronic amplifier. At low input levels, the amplification factor is small; it increases in the middle range and then saturates and approaches a limiting value. The corresponding graph rises slowly, then more quickly, and finally flattens out. This is just one example: nonlinear functions can take almost *any* form, provided there is only one output value for each input (the graph is not allowed to "double back"). A linear function would be represented by a graph that is literally that: a straight line. Hence the world we can describe with linear functions and differential equations is narrow, and we need nonlinear models and methods to deal with many phenomena in nature.

But now suppose that we are interested in the behavior over only a limited range of inputs near some point of interest x_0, perhaps the condition at which the device in question is designed to operate. Then, if the graph does not turn or oscillate violently, we may approximate it with a short segment of the straight line *tangent* to it at x_0. We also show this *linearization of f* in figure 4.8. If the inputs do not depart too far from x_0, remaining between x_1 and x_2, say, then the linearized function provides quite a good approximation to the "true" nonlinear one.

One can use this tactic in studying nonlinear differential equations near equilibrium points. The functions defining the vector field are linearized at the equilibrium and one obtains linear equations, which may be classified as in figure 4.7. Hartman and Grobman showed that, for the full nonlinear system, the structures pictured in figures 4.7a,b,d,e, and f do not change *near the equilibrium*. The qualitative features we are interested in remain the same. The orbits might be curved differently, but they maintain the property of approaching the equilibrium, in case (1) (figs. 4.7a and b), or of having at least one orbit that leaves the equilibrium, in case (3) (figs. 4.7d,e, and f).

Although this idea was suggested by Poincaré as a systematic way to tease qualitative information out of a nonlinear system, it had to await Grobman and Hartman for rigorous proof. It actually seems quite natural, for near the equilibrium the difference between the linear and nonlinear systems is small, as is the difference between the straight line and curved graph of figure 4.8. A small change should not be expected to overturn the robust structures of figures 4.7a,b,d,e, or f. Of course, in cases of instability, such as d,e, and f, once the solution has left the neighborhood of the equilibrium the linearized equations are useless, and we must return to the full nonlinear system. Linearization provides only a good *local* approximation.

In case (2), however, the situation is very different. Under even the smallest perturbation of the differential equation, the orbits may fail to close up and the structure of figure 4.7c can be replaced by either of the forms in figures 4.7b or e. The orbits that formerly circulated without growing or shrinking must still circulate, but now they may gradually spiral in or out. Depending on which way our perturbation nudges the problem, either of the two situations may occur, and stability can become asymptotic stability or instability. In this case we say that the linearization does not determine the behavior of the nonlinear equation; there is, in fact, no foolproof algorithm for determining what happens in this delicate situation.

Liapunov did propose a method of determining stability or instability, but it consists of finding a function with certain properties that mesh nicely with those of the differential equation. The function detects whether the flow crosses a small sphere surrounding the equilibrium in the inward or the outward direction. In the former case, one has stability, and in the latter, instability. Unfortunately, this geometrically simple idea is hard to carry out: there is no simple recipe for finding *Liapunov functions*, so this method is sometimes harder than proving stability or instability in other ways. Nevertheless, no "other ways" as general as the one of Liapunov are presently known.

Another important contribution in Aleksandr Mikhailovich's work was that of defining certain *characteristic numbers* for functions (and in particular for solutions of differential equations), numbers now called *Liapunov exponents*. These exponents quantify the rates of growth or decay of solutions of the linearized equation in various directions in the phase space, generalizing the characteristic exponents of periodic solutions that Poincaré had introduced in his prize memoir. They are of great interest in current research, not only because they determine the stability of individual solutions (when they can be calculated), but because positive exponents, signifying growth, also provide evidence for the presence of chaos.

Liapunov's thesis inaugurated a new field. The problem of stability was separated from the specific study of the solar system and became a part of the theory of differential equations, completely independent of its origin. The advantage of the new method was that it could be applied to *all* differential equations. It therefore addressed a broad variety of questions in many branches of science. The two equilibrium solutions of the pendulum discussed in the first chapter (see fig. 1.7) were now proven rigorously to be respectively stable and unstable. Many more complicated questions have also been settled, and the *theory of stability* has become a subdiscipline in its own right.

In the century since Liapunov's doctoral thesis, his original ideas have been greatly expanded, leading to new concepts and methods. They find application today not only in mechanics and physics, but in chemistry, theoretical biology, and in practically every area of applied science and engineering. The original question of the stability of the solar system has been expanded into a rich interdisciplinary science with ramifications in the most unexpected corners of knowledge. It has also provided the soil in which the roots of *chaos* have grown.

The notion of Liapunov stability refers to a single orbit and the bundle of solutions in its immediate neighborhood. All are solutions to a given, fixed differential equation; only their initial conditions differ. To understand if the flow of a differential equation as a whole has stability properties with respect to external perturbations of the functions defining it, a new idea was needed. This was created in the 1930s in the work of two other Russian mathematicians, A. A. Andronov and L. S. Pontryagin. It is called *structural stability* and occurs as part of *bifurcation theory*, which also has its roots in Poincaré's studies of the restricted three-body problem. In some translations of the Russian literature, structurally stable systems are referred to as "coarse" or "rough" systems (*systèmes grossieres*), and the complementary, structurally unstable or bifurcation systems are called "fine."

In order to understand the complementary concepts of *structural stability* and *bifurcation*, we again appeal to the metaphor of a flow on the surface of a river. This time we imagine that the flow depends on an external *parameter*, a quantity that represents a change coming from outside the system itself. One might think of rainfall in distant mountains at the river's source, or the wind that blows across its surface, influencing the patterns and currents on it. For simplicity we will take the speed of the wind as our only external parameter.

Roughly speaking, the flow is said to be *structurally stable* if its structure remains qualitatively the same, independent of small changes in wind speed. If the structure changes, the wind speeds at which these changes take place are called *bifurcation values*. Let us take two examples.

In figure 4.9a the flow (under still air, say) is formed by parallel lines. As the wind picks up, the lines become slightly curved, but the picture does not change its qualitative structure (fig. 4.9b). By this we mean that no new qualitative feature appears, such as an eddy, for instance. If this is true for *any* small variation of the parameter, then we say the flow is *structurally stable* at the initial value of the parameter: zero wind speed in our case.

We can now imagine all kinds of possible pictures in phase space. For example, suppose that for negative values of the parameter, the flow contains a stable equilibrium point toward which all nearby orbits spiral as in

(a)   (b)

Figure 4.9. A structurally stable flow (a) before and (b) after perturbation. There is no qualitative change in behavior.

figure 4.10a. For positive values, neighboring trajectories spiral away from the equilibrium toward a (small) closed curve that surrounds the now unstable equilibrium, as in figure 4.10b. From farther out, however, orbits still spiral inward. The picture in the large is much the same; the change that has occurred is *local*. The borderline case or *bifurcation point* separating the two portraits is *structurally unstable*.

The qualitative change that has taken place is clear: in case (a) of figure 4.10 the oscillations die out as time progresses; in (b) they settle on the small periodic orbit. At the bifurcation point corresponding to the parameter taking the value zero, the equilibrium is degenerate: its linearization looks like the picture of figure 4.7c, but the nonlinear terms cause a weakly stable spiraling behavior. As the parameter varies, the periodic orbit grows in size. This particular bifurcation bears the name of the German mathematician Eberhard Hopf. It is, in fact, part of a theory developed first by Poincaré and later by Andronov and his collaborators for two-dimensional systems. In 1942, Hopf generalized the theory to differential equations of arbitrarily high dimension. Hopf bifurcations occur in models of many kinds of systems. Two disparate examples are the spontaneous oscillations in electronic and acoustic circuits that lead to the unpleasant squeals of feedback in audio amplifiers, and "nose wheel shimmy" in aircraft landing gear.

(a)   (b)

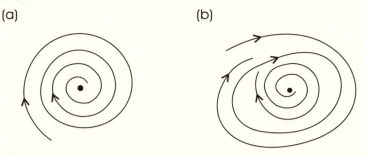

Figure 4.10. The Hopf bifurcation, showing phase portraits (a) before and (b) after the birth of a limit cycle.

The notion of structural stability is important for practical reasons. Structurally stable flows are robust, so modest variations in external parameters will not unduly influence them. If a phenomenon is described by a structurally stable flow, its behavior is resilient and resistant to changes in the enviroment. In the face of change, the whole set of solutions adapts with flexibility, exhibiting only slight modifications that do not destroy its essential character.

Structurally stable flows confer a certain robustness on the mathematical models in which they occur. Measurements are never absolutely exact; one must always take a margin of error into account. If a bifurcation occurs at the value we have measured, we cannot say to which type of behavior the physical situation corresponds. If the flow is structurally stable, then we are able to guarantee qualitative accuracy around the measured data.

Catastrophe theory, a fairly young branch of mathematics that stimulated a number of controversies and some journalistic turbulence during its early years, has adapted the notion of structural stability to its own purposes. In this case also, one is concerned with how the qualitative properties of functions depend upon external parameters.

Structural stability and bifurcation remain central issues in the theory of dynamical systems, with important implications outside of mathematics. They have already found applications in the study of the heart, in embryology, linguistics, optics, psychology, hydrodynamics, economics, elementary particle physics, geology, and numerous other areas.

PLANETS IN BALANCE

An important notion related to stability questions in celestial mechanics is that of a *central configuration*. In 1767 the sixty-year-old Leonhard Euler published a study of the three-body problem. In it he explicitly obtained a whole class of periodic solutions. He proved that if three particles of arbitrary (finite) masses are arranged initially on a line, as in figure 4.11, such that the ratio *AB/BC* has a certain value given by a complicated formula depending on the masses, and if suitable initial velocities are assigned to the particles, then they will move periodically on ellipses, maintaining at all times a collinear configuration. Moreover, the ratio *AB/BC* of distances measured along the rotating line *AC* will remain unchanged throughout the motion. This is quite striking since, as we have pointed out, the addition of a third body to the two-body problem generally perturbs the elliptical Keplerian orbits and may even cause chaos.

In 1772 Lagrange rediscovered the Eulerian solutions of the three-body problem and found a second important class of orbits. He showed that if, at the initial moment, the three particles lie at the vertices of an equilateral

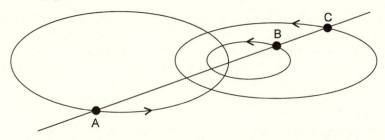

Figure 4.11. The Eulerian solutions of the three-body problem.

triangle (a triangle with equal sides) and if, again, suitable initial velocities are taken, then the masses will move periodically on ellipses, as in figure 4.12, preserving always their equilateral configuration. The triangle will change its size and orientation as the particles orbit, but throughout it remains equilateral. Not surprisingly, these are called *Lagrangean solutions*.

There are three kinds of collinear solutions, corresponding to the different ways of ordering three points on a line, and two of triangular type, corresponding to the two distinct orientations of a triangle. In both cases, if the bodies are arranged in the collinear or equilateral configurations and if the initial velocities are set equal to zero, then the particles will move toward their common center of mass and collapse simultaneously in a triple collision at that point in finite time. This is the limiting case in which the elliptical orbits degenerate to line segments, and the solution is no longer periodic.

In these special solutions, all the mass particles perform an identical dance, so the full nine degrees of freedom available to the three bodies remain unexplored. The motions occur on an invariant submanifold of lower dimension, of the kind introduced in chapter 1. As we noted there,

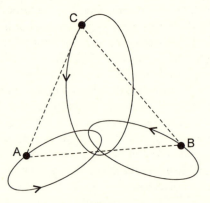

Figure 4.12. The Lagrangean solutions of the three-body problem.

invariant manifolds can form a kind of skeleton that helps us understand the full body of phase space. We also saw in chapter 3 how Xia and Gerver used such symmetries in constructing their examples of noncollision singularities.

To see the connection between stability and these rather special classes of Eulerian and Lagrangean solutions of the three-body problem, let us suppose they move on *circular* orbits and examine these orbits in rotating coordinates. Here we are bound to an observation frame rotating in the same manner as the line or the triangle, within which we do not perceive the motion of the particles, much like riding in a car and watching a train traveling at the same speed on a track parallel to the road. We see three bodies in stationary positions. In both cases, they represent equilibrium positions in the new coordinates, and we can therefore investigate the stability of these equilibria. In fact, for suitably chosen (and more complicated) coordinates, the solutions can be viewed as equilibria even when the orbits are ellipses rather than circles. In either case, such equilibrium solutions are called *central configurations*.

There is another way to look at these solutions. Consider a two-body problem (the motion of the earth on an ellipse around the sun, for example) and suppose a third body of small mass (a spaceship) moves in the gravitational field of the larger bodies. The small body does not influence the motion of the large ones, but its motion is determined by both of them. This is a restricted three-body problem of the type we have met several times already. There are five special positions that the small body can occupy in the plane of the elliptic orbits described by the large masses. They are called *libration points* and are denoted by L_1, L_2, L_3, L_4, and L_5 (see fig. 4.13). The notation is classical, and the letter L was assigned to honor Lagrange. The term "libration" derives from the Latin *libratio*, which means balance. Two different meanings of this word occur in astronomy. Here it signifies that a particle at the libration point is *in balance*, at rest in the rotating frame, the centrifugal forces and gravitational pulls of the two larger bodies just canceling. The word "equilibrium" itself comes from the same root. The other sense describes the *libration of the moon*, the oscillatory motion on which Lagrange also worked. The libration in longitude, as we mentioned earlier in this chapter, alternately reveals and hides the eastern and western borders of the moon's visible face.

If we consider two large masses and place the small one at one of their libration points, we recover the Eulerian and the Lagrangean configurations. In this new setting it is easier to study the stability of libration points instead of the stability of Eulerian-Lagrangean solutions, but of course this works only for a small third mass. We know today under what circumstances the libration points are stable; this property depends on the relative values of the larger masses. An example of a stable libration point in the

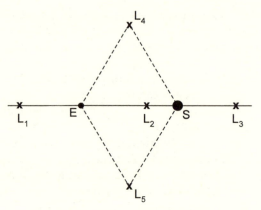

Figure 4.13. The five libration points.

astronomy of the solar system is provided by the Trojans, a cluster of asteroids moving on approximately the same orbit as Jupiter around the sun and
forming at all times an equilateral triangle with these two much larger celestial bodies. Also, at the Lagrangean points of the earth-moon system,
astronomers have found clusters of cosmic dust. It has been suggested that
the stability of these points might be used to make the zones around them
parking places for future space stations or for building cosmic colonies.

As we saw in figure 4.13, there are five classes of central configurations
for the three-body problem: two equilateral (one for each possible orientation of an equilateral triangle with the particles at its vertices) and three
collinear (one for each possible ordering of three particles on a line). Surprisingly, very little is known about the number of classes of central configurations for the n-body problem in general, when n is larger than three. We
do not even know if they are finite or infinitely many, and should it be the
latter, we do not know whether they make up a set of *isolated points* (i.e.,
points having a positive distance between them) or a *continuum* (i.e., a
continuous portion of phase space). A number of people have worked in
this area; the interested reader will find their names in the Notes section
following chapter 5.

In 1912 Karl Sundman discovered an interesting property of the three-
body problem. He showed that if the three particles collide simultaneously,
then they must necessarily tend toward a central configuration. In other
words, before colliding, the particles come very close either to an equilateral triangle or to a collinear configuration of Eulerian type. Thus the invariant manifolds mentioned above provide a sort of highway leading to
triple collisions. This property has been proved to be true in the n-body
problem, in the sense that simultaneous collision of any subset of k particles tends toward the set of central configurations formed by those k par-

ticles. Unfortunately, since the set of central configurations might be a continuum, we still do not know if a collision orbit tends toward a specific central configuration or oscillates among several, without settling on any of them.

Moreover, Donald Saari proved in 1983 that a large class of solutions of the n-body problem that scatter to infinity do this by tending toward central configurations as the bodies move apart. (For example, if three noncollinear particles scatter to infinity, the farther they separate, the closer they approach the corners of an equilateral triangle.) This result sheds some light in the cosmological problem of the motion of galaxies. In spite of its seemingly special nature, the notion of central configurations remains an important subject in contemporary celestial mechanics.

In this and in the preceding chapters, we have traced the beginnings and some of the ramifications of two fundamental concepts in dynamical systems theory: chaos and stability. We saw how Poincaré's search for stability led him to the discovery of chaos. The symbolic dynamical analysis of Smale, described in chapter 2, revealed that such deterministic chaos is in many ways indistinguishable from a truly random process. Its hallmark is sensitive dependence on initial data: chaotic orbits are effectively unpredictable in the long term. In contrast, Liapunov stability, introduced above, typically implies unchanging behavior or regular, predictable repetition. It might by now seem to the reader that these two properties are mutually exclusive, if not utterly opposed. Why, then, have we linked them in the subtitle of this book, if chaos only appeared accidentally, in seeking stability? It was to turn out that they are, indeed, intimately interwoven in a particularly beautiful manner in certain dynamical systems: the nearly integrable Hamiltonian systems of classical mechanics. Our final chapter will be devoted to these problems. We shall meet the triumvirate of Kolmogorov, Arnold, and Moser, the three mathematicians responsible for the theory that links chaos and stability.

5.

KAM Theory

> We may formulate the aim of dynamics as follows: to characterize completely the totality of motions of dynamical systems by their qualitative properties.
> —George David Birkhoff

AMSTERDAM 1954. The final day of the International Congress of Mathematicians was almost over. Only the last plenary lecture remained. The honor of delivering the closing words had been given to the "Euclid of probability theory," the man who had provided an axiomatic basis for problems of gambling and games of chance, connecting them to measure theory: the Russian mathematician Andrei Nikolaevich Kolmogorov.

The topic of the lecture was rather unexpected. Kolmogorov was not going to discuss probabilities. The title announced in the program was "On the General Theory of Dynamical Systems and Classical Mechanics." Relatively few of the audience knew that Kolmogorov had been working in this area over the previous years.

The speaker was introduced and the audience greeted him with applause. "It was a surprise to me," he began,

> that I would have to present a paper at the final session of the Congress in this large hall, which was known to me rather as a place for the performance of the great musical compositions of the world, conducted by Mengelberg. The paper I have prepared, without taking into account that it would occupy such an honored position in the program of the Congress, is devoted to a rather special range of problems. My aim is to elucidate ways of applying basic concepts and results in the modern general metrical and spectral theory of dynamical systems to the study of conservative dynamical systems in classical mechanics. However, it seems to me that the subject I have chosen may also be of broader interest, as one of the examples of the appearance of new, unexpected, and profound relationships among different branches of classical and modern mathematics.
>
> In his famous address at the Congress in 1900, David Hilbert said that the unity of mathematics and the impossibility of its division into independent branches

stem from the very nature of this science. The most convincing evidence for the correctness of this idea is the appearance of new focal points at each stage in the development of mathematics, where, in the solution of quite specific problems, notions and methods from quite different mathematical disciplines become necessary and are involved in new interrelations.

Kolmogorov was a thinker who constructed bridges between different branches of mathematics, who saw relations that no one had seen before. His was a global vision of science, but he also knew how to enter the most intimate details of the problems he tackled. In his introductory remarks he expressed his broad view of mathematics and, at the same time, set the stage for one of the more remarkable mathematical achievements of this century, the creation of KAM theory. The acronym KAM comes from the names of Kolmogorov, Arnold, and Moser, the three mathematicians who laid the foundations of this theory between 1954 and 1963, and to whose work this chapter is devoted.

Simplify and Solve

Andrei Nikolaevich Kolomogorov was born in Tambov, Russia, on 25 April 1903. His father, Nikolai Mateevich Kataev, the son of a priest, was an agronomist. His mother, Maria Yakovlevna Kolmogorova, the daughter of Marshal Yakov Stepanovich Kolmogorov, died in childbirth, and the small boy was raised by her sister Vera. When Andrei was six, he and his aunt moved to Moscow. It was here that he would start school and, with only short breaks, spend his whole life.

Kolmogorov's childhood and teenage years were far from easy. The First World War and the bloody events following the Bolshevik revolution, the terror of the civil war and the famine that followed it, the many tragic consequences of the struggle for power between the Red and the White Russian armies: all were part of the young man's life. In the immediate postrevolutionary period 1919–1920, he worked for the railway to make a living. But in spite of these hardships and disruptions to the normal youth of an academically oriented child, Andrei Nikolaevich managed to enter the physics-mathematics department of Moscow University and to begin his academic studies in 1920.

His first year as a student was marked by the presence of three remarkable Russian mathematicians: N. N. Luzin and A. K. Vlasov, from whom he took courses in function theory and projective geometry, and V. V. Stepanov, who conducted a seminar on trigonometric series. After recognizing Kolmogorov's talent, Luzin approached the new student with a proposal to collaborate on research together.

Plate 5.1. Andrei Nikolaevich Kolmogorov. (Courtesy of Nauka, Moscow)

Kolmogorov had a natural feeling for mathematics, and his skills developed rapidly in this academic environment. In his first year, at the age of eighteen, he completed a study of trigonometric series, and a further year later he published a second paper in descriptive set theory. A third article, dated 2 June 1922, brought him true scientific recognition. In it, he produced the first example of a trigonometric series divergent almost everywhere. This settled an important question in connection with Fourier series, which are useful not only in pure analysis but also in many areas of applied mathematics. In fact, the subtleties of Fourier series of different sorts of functions and their rates of convergence were to play a crucial role in the KAM theory some forty years later. After graduating in 1925, Kolmogorov began his work in probability theory, and within a few years he had established his reputation.

Kolmogorov's intellectual interests and contributions were not restricted to mathematics. Before his debut in the exact sciences he wrote a historical study of the medieval Russian republic of Novgorod. Then, on learning that his professors demanded several arguments to establish each claim, he decided that mathematics was easier, since a theorem required only one proof. However, he did not abandon his many other interests; rather, they expanded to encompass a multitude of scholarly fields including genetics, poetics (the theory and practice of poetry), and stratigraphy (a branch of geology dealing with the arrangement of rocks in layers, or strata). A similar wealth of interests is apparent in his mathematical work. He made fundamental contributions not only in probability theory and classical and celestial mechanics, but to ostensibly unrelated branches such as topology, mathematical logic, and algorithmic complexity theory. The three volumes of his selected papers are entitled *Mathematics and Mechanics, Probability Theory and Mathematical Statistics*, and *Information Theory and the Theory of Algorithms*. It is rare to see such a broad range of talents in a single mind.

In order to understand the revolutionary ideas that Kolmogorov introduced at the congress in Amsterdam, we must first present some of the notions on which he drew and which we will use in our description. The first is that of a *Hamiltonian system* of differential equations, or the *equations of dynamics*, as Poincaré called them in both his prize memoir and in *Les méthodes nouvelles*.

This notion is fundamental to the study of dynamics and of mechanics in general, having its roots, like so much else of more general use, in research done on the three-body problem. Actually, the name "Hamiltonian" is not justified historically, since equations of the same form had already been considered by Lagrange and Poisson many years before Hamilton. (For that matter, as Clifford Truesdell has remarked, the first person to write down "Newton's equations" in the form we now know them was not Newton, but Leonhard Euler.) Nonetheless, the Irish mathematician William Rowan Hamilton's contributions to mechanics are now honored by this terminology. Perhaps appropriately, it seems to have been adopted only among mathematicians and physicists, for the same sets of equations still bear the older term *equations of dynamics* among other circles of applied scientists who use them marginally in their research. Hamilton is better known to most mathematicians for his invention of quaternions—an extension of the idea of complex numbers—and for his contributions to algebra. He also developed a unified approach to problems of geometrical optics and dynamics via the *calculus of variations*.

The main ingredient of Hamiltonian mechanics consists in the expression of the equations of motion in an elegant symmetrical form, in terms (of the partial derivatives) of a function called the Hamiltonian. In the original Newtonian formulation, the primary state variables are the position coordinates of the bodies. As we saw in chapter 1, the velocities, needed to describe the full state, are defined "secondarily" as the rates of change of the positions. In Hamilton's equations, velocities are replaced by *momenta*: the products of the velocities and masses of the bodies. Positions and momenta are placed on an equal footing. They both appear in the Hamiltonian function, sometimes simply called the *Hamiltonian*, which is in general the total energy of the mechanical system under study. As such, it provides a relation between the variables involved in the differential equation that will be useful to us later in reducing the dimension of phase space. Different problems correspond to different Hamiltonians. One Hamiltonian describes the *n*-body problem, another describes the motion of a pendulum, a third corresponds to "ideal" (inviscid) fluid dynamics, and so forth. Once one has a theory in which *arbitrary* Hamiltonians are considered, the results obtained can be applied to all the specific problems this theory comprises: to the problems of dynamics.

It is a fundamental tendency of mathematics always to strive to raise concepts to higher levels of abstraction. To understand the reason for this, let us pose the following simple question. Which is easier to compute: two apples plus three apples make five apples; two chairs plus three chairs make five chairs; or simply $2 + 3 = 5$? The answer is obvious. Once the abstraction has been made, the purely numerical statement is easier to manipulate. One has stripped away redundant information (Were the apples ripe? Did the chairs have arms?). The formula $2 + 3 = 5$ applies to all possible objects anyone could ever think of. The symbolic operation is done at a higher level of abstraction. We are so used to working with numbers that we no longer remark the importance and benefits of this process. But humankind needed hundreds of thousands of years to realize this step.

Historical studies in linguistics show that primitive populations typically had one notion for *one tree*, a different notion for *two trees*, and a third notion for *three or more trees*. At that stage, people had not identified the concept of a *numeral*. They did not understand that trees, as well as fruits, stones, or any other objects can share the common feature of being grouped *one, two, three*, or *more*. The same happened with the notion of *leaf*. Some contemporary native peoples have a word for each particular kind of leaf, but lack the notion of *leaf* itself. In a practical sense they are entirely correct. An abstract leaf does not exist in nature; only specific kinds of leaves—oak, elm, birch, ash—can be found. *Leaf* is an invention of our minds: an abstraction.

These examples demonstrate the important role of abstraction in our thinking. Jorge Luis Borges tells a story of Funes, a young man cursed with a perfect memory. Abstraction was unknown and unneccessary for Funes, since he could recall precisely all the details of the appearance of the sky and clouds at dawn on a particular day, or the individual smell and taste of a fruit eaten years ago. Eventually he dies, still young, completely crushed by the weight of his memories. It is so much easier to abandon the colorful, irrelevant details and retain and manipulate only essential notions. One might say that progress, or at least scientific progress, relies on the ability to identify and then ignore irrelevant information in a kind of creative amnesia. The same approach has been used steadily and successfully in mathematics, and Hamilton's equations represent only one among a host of examples. Of course, we must not confuse such simplification and idealization with complete understanding, and the reader should remember that we are concerned only with making and analyzing *mathematical models* of physical phenomena. The model is not reality, no matter how much it helps us come to terms with the facts we originally set out to understand.

Once one has an inclusive, abstract concept such as *leaf*, it is useful to further subdivide and categorize, perhaps in this example into evergreen and deciduous leaves. Returning to Hamiltonian systems, we traditionally divide them into *integrable* (also called *completely integrable*) and *nonintegrable* cases. Integrable systems are "soluble" in the sense that their general solutions are explicitly obtained as formulas involving known functions. They are those for which Newton's anagram of chapter 1, claiming that "the flow can be determined," is practically realized. Although there are infinitely many systems of this kind, they are nonetheless very rare, because they occupy a set of measure zero in the space of *all* Hamiltonian systems. They represent the exceptions. In spite of this, almost all systems described in textbooks are integrable, for they are the only easy ones to explain and about which homework problems can be set. Nonintegrable systems are those for which a solution cannot be found using standard methods. In spite of the optimistic textbooks, almost all Hamiltonian systems are of this latter, messy sort.

The main question that Kolmogorov addressed in Amsterdam was: What happens to solutions of an integrable system when the equations are slightly changed by a small perturbation that respects the Hamiltonian structure? He posed this question in connection with the behavior of a particular class of orbits, which had occupied the minds of Dirichlet, Weierstrass, and Poincaré for more than a century: the so-called *quasi-periodic solutions*. To understand what they are, and to better describe both the question and the answer that KAM theory gave to it, we will again turn to the dynamics of the solar system.

Quasi-periodic Motions[*]

The First Law of Kepler states that the planets move around the sun on elliptical orbits. The German mathematician drew this conclusion after many years of observation, computation, and comparison. Strictly speaking, he was mistaken. His instruments were too crude to provide him sufficiently accurate estimates to see that he had found only a first approximation of the path of Mars, the planet from which most of his data derived. Observations on Mercury or Venus would have more readily shown that their orbits are in fact slowly *precessional ellipses*: orbits whose shapes barely deviate from the elliptical on each circuit, but whose properties gradually evolve over time. We illustrate this in figure 5.1.

Using the gravitational law of Newton, as sketched in chapter 1, one can prove that the motion of a single planet around the sun should indeed be an ellipse, provided that the influence of other planets is ignored. But this influence is not negligible, and the perturbation it creates leads to a precessional ellipse. From this standpoint, the inadequacies of Kepler's instruments were fortuitous. Had he discovered the true complexity of planetary motion, no such simple law might have emerged, and the only fully soluble *n*-body problem of Newtonian mechanics would have lacked such a lovely illustration. Luckily, observational accuracy developed hand in hand with perturbation theory, the beginnings of which we traced in chapter 4, in the work of Laplace and Lagrange. As Ivars Peterson has remarked in *Newton's Clock*, we are fortunate that the solar system is just simple enough for first approximations to work well, and yet sufficiently complicated to tempt us to go beyond them.

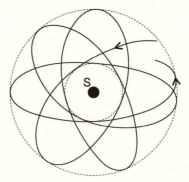

Figure 5.1. The orbit of a planet around the sun, with precession greatly exaggerated.

In the solar system, planetary deviation from elliptical orbits is extremely small. Mercury exhibits the highest departure in precession: the tiny value of just forty-three seconds of arc per century. This means that after hundreds of revolutions around the sun, the *perihelion* of the planet (the point nearest to the sun on the path of a heavenly body) advances by less than one minute of arc. It is this smallness of deviation that makes perturbation theory such a powerful tool in planetary studies. In this chapter, we shall see how Kolmogorov, Arnold, and Moser followed Laplace and Poincaré in perturbing an integrable system to probe the dynamics of an insoluble, nonintegrable one nearby.

Two situations can occur in a precessional ellipse such as the one in figure 5.1: either the curve closes after a while, so the orbit is in fact periodic, albeit with a period of many "years," or it never closes. In the latter case, the path winds densely around inside the ring bounded by the dotted circles. As described in chapter 2, density means that if we follow the orbit for long enough, we can come as close as we wish to any point in the ring. The orbit describing this dense motion is called *quasi-periodic*. A quasi-periodic orbit just fails to become periodic. It repeatedly passes infinitesimally close to the position it had before starting some previous rotation, but the path never actually repeats itself. This provides an example of a dense curve, as promised in chapter 2.

The remainder of this section and the next one are again somewhat technical, as we attempt to describe the structure of invariant tori and the solutions winding about them. The reader may prefer to skim on a first reading and pick up the story at "Letters, a Lost Solution, and Politics."

Let us now imagine how such a quasi-periodic orbit might appear, not inside an annulus in the physical configuration space—here the plane of motion of the planet—but in phase space. In general, this is of high dimen-

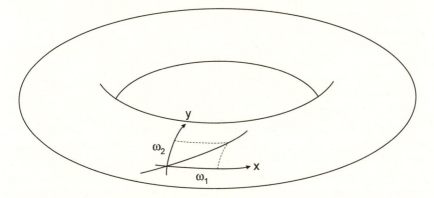

Figure 5.2. A quasi-periodic orbit on the torus fills the surface densely. The coordinates we choose on the torus are *x* and *y*.

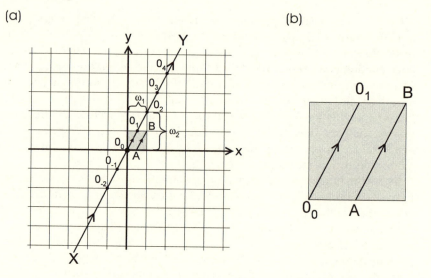

Figure 5.3. Identification of (a) the flow in a plane, with (b) a flow on a square.

sion, for it contains velocity or momentum as well as position coordinates of all the bodies in the system, as we described in chapter 1. But, much as we have before, we can take out the manifold on which the quasi-periodic motion lives and examine it in isolation. In the simplest case, it is a two-dimensional *torus*: the surface of a bagel or doughnut. We will describe the motion by taking coordinates on the torus like those in figure 5.2. At every point the orbit is a line with a slope given by the ratio of two numbers: ω_2/ω_1. The physical significance of ω_2/ω_1 is that it represents a *frequency ratio* relating the two periods in the planetary motion: its basic orbital period and the slow precession of its perihelion. If this ratio is a rational number (a fraction), then the orbit closes and is periodic; otherwise it winds around for ever and fills the torus densely.

To see this we must first understand how a flow given by parallel lines of slope ω_2/ω_1 in the plane can be identified with a flow on a square, and then how the flow on this square can be identified with a flow on a torus. The first step is shown in figure 5.3, the second step in figure 5.4. Let us describe them in detail.

Define a flow in the plane with all orbits parallel to *XY*: a straight line with slope ω_2/ω_1, see figure 5.3a. This is a uniform *translation*—a steady parallel flow like that on the surface of an idealized, straight river. Suppose the ratio ω_2/ω_1 is 2/1. The particular orbit *XY*, of slope 2, cuts the network of squares in the points . . . O_{-2}, O_{-1}, O_0, O_1, O_2, . . . , as shown. We now wish to "compress" this infinite orbit *XY* (only a part of which is shown anyway) into a single square of unit length, say the shaded one in figure

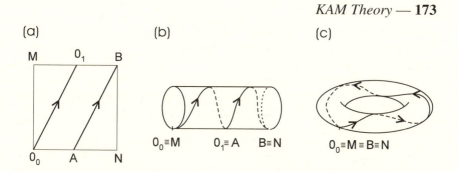

Figure 5.4. The equivalence between the flow on the square and the flow on the torus, obtained by identifying the opposite sides of the square.

5.3a. XY crosses this square in the segment O_0, O_1, starting at bottom left and exiting at top center. It crosses the next square in O_1, O_2, from bottom center to top right. In the following square, O_2, O_3 exactly repeats the route O_0, O_1 through the shaded square (bottom left to top center), and in the next, O_3, O_4 repeats the path O_1, O_2 in the second square. If we look closely at *all* squares crossed by the orbit XY, we see only these two kinds of alternating behavior. So we may represent the whole orbit inside the single shaded square instead of drawing the whole plane: all we have to do is slide the second square over the first (shaded) one and copy the path; this captures all the information. Figure 5.3b shows how the orbit XY is now represented inside the shaded square. We have labeled the "ends" of the second orbit segment as A and B.

We seem to have left the torus far behind, but we are about to find it again. By pasting together the sides MB and O_0N of the square in figure 5.4a, we obtain a cylinder. The points M, O_1, and B are *identified* with O_0, A, and N, respectively. This turns the two orbit segments $O_0 O_1$ and AB into a single spiral that encircles the cylinder twice, as drawn in figure 5.4b. Finally we bend the cylinder and paste the circular ends together, to obtain a torus having the four points O_0, B, M, and N identified. Now the spiral on the cylinder has become a closed spiral on the torus. It closes up because it started through the square at O_0 (bottom left) and exited at B (top right), and all four corners of the square have been identified into a single point by our pasting process. This shows that, if the slope is $\omega_2/\omega_1 = 2/1$, the corresponding curve on the torus gives rise to a periodic orbit. The same thing happens if the slope is *any* rational number, but then the spiral may wind many times around the torus before it finally closes.

In fact, the geometrical manner in which the orbit winds around the torus precisely reflects this rational number. In the example above, $\omega_2/\omega_1 = 2/1$, or equivalently $\omega_1/\omega_2 = 1/2$, the orbit makes one full circuit (longitudinally) around the hole in the doughnut, and two circuits (meridionally) through the hole. In general, for a rational frequency ratio $\omega_1/\omega_2 = m/n$, it makes m longitudinal and n meridional circuits before closing up. Try wrapping a piece of string around a doughnut in this fashion and joining the ends (say, for $m = 2$, $n = 3$). Eat the doughnut (but not the string), and, as long as both n and m are bigger than one and have no common divisors, you will find that your soggy piece of string is tied in a knot. The knot has nothing directly to do with our story, but it is an interesting sidelight on topology. By wrapping up our orbit around a torus we have changed a simple straight line into a knot!

The situation is utterly different if the ratio ω_1/ω_2 is an irrational number (i.e. one which cannot be written as a fraction). In this case the orbit on the torus cannot close up, for if it starts at the bottom left (the coordinate $(0, 0)$, the origin of the plane), it never passes through the corner of any other square. To do so, it would have to climb by some integer number n while advancing m squares to the right. Since its slope is irrational, it cannot do this. It therefore never precisely repeats its path through the shaded square of figure 5.3, so, to represent the entire orbit, infinitely many different orbit segments must be copied. This essentially fills up the whole square. In turn, each path will fill the torus densely. In this case we have a *quasi-periodic* orbit on the torus.

PERTURBING THE TORI[*]

The central issue addressed by Kolmogorov in his Amsterdam talk concerned perturbation of invariant tori. To see why this is a natural problem, we must explain a remarkable property of integrable Hamiltonian systems. If expressed in suitable coordinates, *almost every* solution lies on some member of a family of nested tori, as figure 5.5 suggests.

As we pointed out in chapter 1, for an n-degree-of-freedom Hamiltonian system, the phase space is $2n$-dimensional, and so the two-dimensional tori shown in figure 5.5 can generally be taken only metaphorically. But there is a situation in which this picture is literally correct: that of a two-degree-of-freedom system that conserves energy. Here the reader should recall our discussion of constants of motion and reduction of dimension in chapter 1. Let us denote the two position and two velocity or momentum coordinates by q_1, q_2 and p_1, p_2, respectively. The Hamiltonian—in this case the total energy—is a function of these four time-dependent variables, which we may write as $H = H(q_1, q_2, p_1, p_2)$. H is also called an *integral of motion*.

Now H remains constant as the solution evolves: $H = H(q_1, q_2, p_1, p_2) = h =$ constant. This equation implies that a fixed relationship holds among the four coordinates, even though they themselves change with time, so

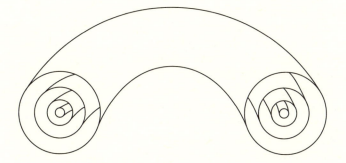

Figure 5.5. A family of nested tori that foliate the energy manifold.

that once any three are specified, the fourth may be derived. Thus, once initial conditions are chosen, the solution of Hamilton's equations remains confined to a three-dimensional *energy manifold*, which locally resembles a piece of our usual three-dimensional space. In this case, figure 5.5 exactly describes the structure of each energy manifold for an integrable system.

The first constant of motion, the total energy $H = h$, has allowed us to reduce the dimension by one, from four to three, by selecting a particular energy level. The fact that the resulting three-dimensional energy manifold is filled with nested two-dimensional tori corresponds to the existence of a second integral that we may write as $K = K(q_1, q_2, p_1, p_2) = k$, in addition to H. For each fixed h we get a picture like figure 5.5. Fixing k additionally specifies another relationship among the four variables, or, effectively, a relation among the three remaining after using H. Each value of k picks out a single one among the continuous family of tori. For this, the two integrals H and K must be *independent*, so that they provide two pieces of separate information about the system.

Restricted to a torus, the behavior of solutions is, as we have seen, simple: they are either all periodic orbits characterized by a rational frequency ratio m/n, or quasi-periodic orbits with an irrational ratio. Since two quantities or *parameters*, h and k, must be set to define a particular torus, we speak of a *two-parameter family of tori*. Every point in phase space lies on some torus in some energy manifold. We say that the tori *foliate* the manifold, rather as the leaves or pages make up a book. This is what is meant by complete integrability for a two-degree-of-freedom system. As we saw in chapter 1, one of Poincaré's major achievements in his prize-winning paper was to prove that additional integrals of a certain type do *not* exist in the restricted three-body problem.

We spoke of a *frequency ratio*. Roughly speaking, an integrable two-degree-of-freedom system can display two kinds of motion, each with its own characteristic period or time scale. When the two are related by a rational number, we say that the resulting periodic orbits lie on a *resonant* torus, and when the relation is irrational, a *nonresonant* one. It usually happens that, as one changes the parameters h or k (choosing different energy levels, for instance), the corresponding tori change from resonant to nonresonant.

What may be harder to believe, but is particularly beautiful, is that the notion of integrability extends to any number n of degrees of freedom. (It even extends to infinitely many degrees of freedom in some cases.) Provided there exist n-independent constants of motion, one can find n-parameter families of n-dimensional tori filling up phase space, and on each torus the flow is periodic or quasi-periodic.

Returning to the two-degree-of-freedom case, it can be shown that, provided the frequency ratio ω_1/ω_2 changes in a nondegenerate way as one varies the parameters h and k, the set of nonresonant tori has positive Lebesgue measure. (It is difficult to define *nondegenerate* without getting into technical details; roughly speaking it means that the frequency ratio should increase or decrease smoothly as one "moves out" across the tori, increasing k and h. Most Hamiltonian systems are nondegenerate in this sense.) On the other hand, the set of resonant tori has measure zero, but is dense in the set of all tori. These sets of tori are not separate but intimately interleaved so that next to every irrational torus there is a rational one, and vice versa. The situation is precisely analogous to that of the sets of rational and irrational numbers on the real number line. Although both sets are dense, the former has zero measure while the latter has positive measure.

What happens if one takes a completely integrable two-degree-of-freedom system and perturbs it just a little? Will all the invariant tori be destroyed? Intuition suggests that this is likely. The Hamiltonian will change, although it will still keep solutions on a three-dimensional manifold. However, the perturbed system may not remain completely integrable, and if no function analogous to K exists, there is nothing to further restrict solutions to two-dimensional manifolds. So the tori might be expected to break, allowing solutions to wander freely in the three-dimensional energy manifold. Surprisingly, Kolmogorov claimed exactly the reverse. He stated that *most of the nonresonant tori will suffer slight deformation, but will not break apart if the perturbation is small enough.*

The meaning of "most" in the sentence above is that all the resonant and some of the nonresonant tori (forming a set of positive measure) will indeed be destroyed by the slightest perturbation, but this set is nonetheless small in comparison with that of the surviving nonresonant tori. Moreover, as the perturbation size is taken smaller and smaller, the set of nonresonant tori that do break up also becomes smaller and smaller. As the example of a "thick" Cantor set of positive Lebesgue measure in chapter 3 showed, the broken tori are analogous to those points in the gaps removed from the set, while the surviving tori correspond to the points remaining in the Cantor set itself. Around each survivor, there are the remains of broken tori, but the set of survivors nonetheless has positive measure.

Although Kolmogorov published the outline of a proof of his claim in the same year, 1954, he never supplied the full details. It would take a further eight years for Moser and Arnold to publish complete, rigorous arguments, independently and in different settings. In the introduction to one of his own papers, Arnold characterizes Kolmogorov's idea as ". . . simple and novel . . . , the combination of very classical and essentially modern methods, the solution of a 200-year-old problem, a clear geometrical picture and a great breadth of outlook—these are the merits of the work. Its deficiency has been that complete proofs have never been published."

Before discussing the contributions of these two mathematicians, let us try to understand the deeper meaning and implications of the result announced in Amsterdam. Where does this problem come from? Why is it important?

Letters, a Lost Solution, and Politics

Sonja Kovalevskaia was scarcely twenty in 1870 when she met Karl Weierstrass. At fifty-five, Weierstrass was considered the greatest analyst of his time. He held a chair in mathematics at the Royal Polytechnical School in Berlin, to which Sonja had come to study from her native Moscow. In her attempt to enter what was at that time an entirely male profession, she had to overcome considerable obstacles, not the least of which was being barred from attending certain classes. Academic life was ill-prepared for the liberation of women. In Göttingen, for example, when the first woman was to be awarded a doctoral degree, some members of the university senate objected that "she might also want to become a professor and then have the nerve to stand as a candidate for the senate!" In an attempt to moderate his colleagues' opposition, the mathematician David Hilbert responded: "But my dear gentlemen, the senate is not a Turkish bath [*Badeanstalt*]!"

Due to Weierstrass's ill health, he had to present his lectures seated, while an advanced student wrote his notes on the blackboard. Nonethless, he was regarded as a master expositor and his classes were crowded. Kovalevskaia was not permitted to attend his public lectures, but Weierstrass soon heard of this determined and eager young woman and arranged to tutor her privately. They met every Sunday afternoon at his home, and once a week he returned the visit at her lodgings. The friendship begun in this way was to last over two decades and would end only in the painful and premature death of this remarkable woman.

Plate 5.2. Sonja Kovalevskaia. (Courtesy of Mittag-Leffler Institute)

Some years after graduation Kovalevskaia was appointed to a professorship in Stockholm, where she carried out research on the rotation of a rigid body about a fixed point. In general, the equations of motion of such a body have three degrees of freedom, the orientation of the body being specified by three angles: a latitude, a longitude, and a spin angle. Only two completely integrable cases were known at that time. In the first—corresponding to a body freely tumbling in space without feeling gravity—three components of angular momentum and energy are conserved: this leads to the so-called *Euler equations*. In the second, the body has the symmetry of a child's spinning top, being a solid of revolution such as one might turn on a lathe. Here the symmetry is responsible for the conservation of two components of angular momentum, and, with energy conservation, the system is still integrable: this is called the *Lagrange top*.

Kovalevskaia found a third completely integrable case, in which the moments of inertia of the body are related in a particular way. This *Kovalevskaia top* is still regarded as too difficult for inclusion in most textbooks, a fact that hints at her brilliance and creativity. Her article on it won the Borodin Prize of the French Academy of Sciences in 1888, on which occasion the committee doubled the usual prize money, in recognition of an unusual achievement. She became the leading woman mathematician at the beginning of the modern era, earning the respect and confidence of many of her male peers. Her colleague Gösta Mittag-Leffler sought her advice and help in his attempts to set up the King Oscar Prize jury. In addition to her mathematical work, she also wrote a novel, *Vera Vorontzoff*, concerning her experiences as a girl in Russia.

In spite of Kovalevskaia's important contributions to mechanics, we have recalled the story of Sonja Kovalevskaia and Karl Weierstrass for another reason. The friendship between these two unusual people led to a long correspondence, and the roots of many mathematical ideas can be found in their letters. In particular, reading those of Weierstrass, one begins to understand the way in which he thought and worked. It is very different from going through his mathematical papers. Scientific articles, especially mathematical ones, present only the finished products. They conceal the substance of creation, the journey toward the final result, the hopes and the doubt, the passion, the false starts and mistakes, the disappointments. This is exactly what the letters contain. They delineate the human mind behind the abstract theorems. They show us the unsuccessful attempts that were never published. They allow us to follow the halting process of invention.

In these letters we may trace the origins of our question. On 15 August 1878 Weierstrass wrote that he had found formal series expansions for quasi-periodic solutions of the n-body problem and was struggling to prove their convergence. Unfortunately all of his attempts, thus far, had failed. Nevertheless, he continued to believe that these series do converge. There were two reasons for his conviction. One was purely intuitive, a feeling acquired after years of experience that told him, with no clear logical foundation, how things should be. But there was a second reason. Another influential mathematician had left some evidence in this regard.

In 1858 Lejeune Dirichlet told his student Leopold Kronecker that he had discovered a new general method for tackling and solving all the problems of mechanics. Dirichlet had been Gauss's successor in Göttingen. He was known as a trustworthy person and a highly respected mathematician. His work, primarily in number theory, is a model of rigor and correctness. His grand claim had therefore to be taken seriously. Unfortunately, Dirichlet died shortly after making this statement, without leaving any clearly written documentation. Kronecker and others suspected that the method must be connected with series approximations to solutions of the type that Weierstrass would consider later, and that Dirichlet had found a tool for proving their convergence.

We can now more fully understand why Weierstrass proposed the problem of finding convergent power series solutions of the n-body problem for the prize established by King Oscar, as described in chapter 1. Having been unsuccessful himself, but realizing the value that such a method would have for mathematics, Weierstrass hoped to stimulate young researchers to pursue the problem and in doing so recover the method of Dirichlet. He knew that many unanswered questions would benefit from the rediscovery of this method, especially those concerning the stability of the solar system. As we saw in our story of the discovery of chaos in chapter 1, his confidence was amply repaid, although not in the manner he expected.

Over a century after it had first been posed, Kolmogorov was now essentially claiming to have solved this fundamental problem. However, he did not know about the unpublished ideas of Weierstrass and Dirichlet, and his route to the problem had been entirely different. The seeds of his interest had been sown during childhood when he read Flammarion's astronomy book. His dream of understanding one of the great secrets of the universe revived several decades later when, as an established researcher, he found two new sources of information. One was John von Neumann's spectral theory of dynamical systems, included in a book written in German in 1932, *The Mathematical Foundations of Quantum Mechanics*; the other, a paper by Nikolai Krylov and Nikolai Bogolyubov, from Kiev, published in 1937 in the *Annals of Mathematics*, which applied general notions and methods from measure theory to dynamical systems.

The original question that Kolmogorov addressed concerned which ergodic sets, in the sense defined by Krylov and Bogolyubov, actually exist in the flows of the differential equations describing classical mechanics, and which of them have positive measure. Roughly speaking, an *ergodic set* is a part of a flow that cannot be decomposed from the dynamical point of view. It forms an entity in itself, and any attempt to divide it and study its parts in isolation fails. It is inextricably stitched together, usually by means of a dense orbit. The invariant Cantor set Λ of Smale's horseshoe in chapter 2 is an example. If an ergodic set has positive measure, then its presence becomes significant and it cannot be ignored.

We shall say no more on how this specific question translates into mathematical terms, but we remark that it went well beyond the invariant tori problem. As part of his attack on it, Kolmogorov began to study perturbations of integrable systems and discovered the persistence of invariant tori. The original question still remains unanswered today, but this is irrelevant; there is little interest in it now. The results Kolmogorov obtained became much more important than the one he sought initially. V. I. Arnold, who was Kolmogorov's student, has remarked that this is similar to Columbus's discovery of the Americas, when his original goal was to find a western route to India.

After the Amsterdam congress, Kolmogorov's interests turned for a time to *dimension theory* and the *capacity of sets*, but he returned to dynamical systems in the late fifties. In 1957–1958, he taught a course in Moscow in which he discussed the tori results. He also ran a seminar at which physicists and mathematicians spoke on various problems, including magnetic surfaces and the confinement of plasmas. In that application, preservation of tori implies the integrity of the "magnetic bottle" in which a fusion reaction is supposed to be confined. Kolmogorov himself lectured on the three-body problem, and it is clear that by this time he and his colleagues were well aware of the important physical implications of the theorem.

These were some of the scientific motivations for Kolmogorov's interest in the problem. But there were other, larger influences at work in the Soviet Union at this time. Kolmogorov's life had not been easy. He had survived difficult childhood and teenage years, and, while he had been able to travel in the late 1920s and early 1930s, spending periods in Göttingen and France, much of his life had been spent in the shadow of Stalin's great terror, a period in which over twenty million Soviet citizens lost their lives. In 1940 he had himself come close to provoking the regime after publication of a paper on the statistical justification of Mendelian heredity laws. This aroused the ire of the influential geneticist, T. D. Lysenko, whose spurious theories had gained official sanction. Lysenko and a colleague, the mathematician E. Kolman, denounced Kolmogorov's work in papers in which they claimed that "biological regularities do not resemble mathematical laws" and adduced the support of Engels' and Lenin's writings. Nonetheless, as a member of the Academy of Sciences since 1939, Kolmogorov had been more fortunate than some. (The nature of this good fortune may be understood from the fact that, during the Second World War, Stalin authorized the issue of one blanket for each academician, since the temperature within the Academy building in the winter months remained at freezing.)

In 1953 the death of the dictator gave Kolmogorov for the first time a feeling of hope. He and some of his colleagues regained the ability to travel abroad; in fact, he was one of a Soviet delegation of four at the Congress in Amsterdam with which we opened this chapter. It was possible once more to contact foreign scientists and to live without the fear of being sent to one of Siberia's work camps. Although restrictive and petty when compared with the rights and freedoms of Western citizens, Khrushchev's "opening" represented a significant step for Soviet intellectuals. The influence on Kolmogorov's life was profound. He was able to attend subsequent meetings, such as the International Congress held in Stockholm in 1962, and even to spend a semester in Paris. The ten-year period following 1953 proved to be the most productive of his life.

In this way Andrei Nikolaevich Kolmogorov raised the question of invariant tori and stated his theorem on their persistence under perturbation. Using different terminology and new methods, he returned to the problem considered by Dirichlet and Weierstrass. But was Kolmogorov's sketch of the proof adequate? Was such a long-standing problem solved at once, by a single person in a lone battle? That would have been too neat to be true.

Worrying at the Proof

Jürgen Moser read through the proof again and again, trying to understand all the unwritten details. He was wondering why they were not more clearly stated in the article, since he did not find them at all obvious. An editor from the *Mathematical Reviews*, a journal that provides a useful service by printing brief summaries of technical articles, had asked him to produce a short account of a paper. Moser was finding it hard to complete this task, which under normal circumstances is a fairly routine matter. He would have preferred to avoid the struggle that was going on within himself, but, once started, he knew he would have to continue to the end.

The article on his desk was the published version of Kolmogorov's Amsterdam talk. In the course of preparing his review, Moser had consulted several other papers referenced by the Russian mathematician. The missing details of the invariant tori theorem were supposed to be in one of them, but he could not convince himself that Kolmogorov's argument was complete. The proof of convergence of a certain series representation, the core of the edifice, was missing. He thought at first that his own reasoning was wrong, that the statement was most likely obvious, and it was simply that *he* had failed to penetrate it. This made him even more eager to understand. He tried again and again but without success. After several days he realized that nothing was clear. All his attempts to gain insight led to new complications. They seemed insurmountable at first glimpse. They still appeared hard after the most ingenious attempts to overcome them.

On the other hand, Kolmogorov was widely known and respected, while Moser, still in his twenties and not long past his Ph.D., had as yet no reputation. But his conscience drove him to get to the bottom of this problem. It would have been easy to let his doubts be swayed by the fame of the great Russian, to write a laudatory review and to go back to his own research. However, he did not succumb to that easy temptation. After convincing himself that there was no obvious proof of convergence, Moser wrote his summary, published in one of the 1959 issues of the *Mathematical Reviews*, in which he noted that the convergence discussion did not convince him. He ended the review by praising the result that would follow from Kolmogorov's article, if the convergence question were really settled.

The problem continued to bother Moser. He did not consider it closed after sending his review to the journal. He wanted to understand if Kolmogorov's claim was correct or not. His curiosity about the status of the result and its proof superseded his interest in all the other problems he was working on. He became almost obsessed by it.

One day, on the way to his office at the Massachusetts Institute of Technology in Cambridge, Moser suddenly remembered a discussion that had taken place six or seven years earlier. At that time he was still in Germany at the University of Göttingen, shortly before leaving Europe for the United States to take up a Fulbright fellowship at New York University. At the end of a seminar he had approached the speaker, Carl Ludwig Siegel, to ask him some additional questions about a subject touched upon during the lecture.

Siegel was one of the leading mathematicians in the world during those years. Like Kolmogorov, he had made crucial contributions in branches of the subject that ostensibly had nothing to do with each other, and his results are still frequently cited in fields such as number theory and celestial mechanics. He had been one of the few German scientists who returned to that country after the Second World War.

Jürgen Moser himself had come to Göttingen in 1947, at the age of nineteen, leaving East Prussia's Königsberg occupied by Soviet troops. Communist rule had been imposed on the former German region, and the name of this beautiful Baltic city, where Euler had lived and worked for many years, was now Kaliningrad. Moser seemed to have made the right choice in immigrating to the Federal Republic of Germany. He began by writing his dissertation under the direction of Franz Rellich, but in 1950 Siegel returned to Göttingen from Princeton, and Moser audited his lectures on number theory. The young man was deeply impressed by the powerful mathematics, the high standards, and by the rigorous style Siegel promoted. Subsequently, Rellich helped arrange for Moser to write up the notes on Siegel's lectures on celestial mechanics, which would first appear in German in 1956, the year after Rellich's premature death. (The book was dedicated to the memory of Rellich.) Thus, while Moser's interests moved away from those of his first mentor, he owes a considerable debt to Franz Rellich's insight and generosity. For his part, Siegel quickly recognized the talent of this new student, and they began to collaborate as early as 1951, while Moser was still finishing his thesis under Rellich's direction.

The discussion Moser remembered having had with Siegel in Göttingen

Plate 5.3. Jürgen Moser. (Courtesy of J. Moser)

concerned the same type of problem that had now been treated by Kolmogorov. Siegel had said something like "this question seems very important to me," but at that time Moser was not prepared to understand the implications of the statement. Now it occurred to him that this must be the reason he had become so involved. Subconsciously recognizing the similarity of the two questions, he had recalled Siegel's remark, and this had stimulated him to pursue the issue.

Moser trusted Siegel's judgment on the problem's importance, and now another authority in the field, Kolmogorov, had tackled the subject and gone even farther. This persuaded him that the problem must be worth studying. Yet in spite of this assurance, he still wanted to know *why* this was the case: what had made these people decide that the problem was important? In any case, his mathematical taste and curiosity were insisting that he must resolve the question of invariant tori, regardless of its importance. Over the next three years he would succeed in doing both: solving the problem and comprehending its enormous significance.

In the meantime, Moser returned to the Courant Institute at New York University, where he had first held his Fulbright fellowship. This time he had the security of a permanent job, so he could concentrate fully on the problem of invariant tori. He talked to several other people about it, and he even met Siegel a few times, as his former supervisor traveled back and forth between Europe and North America. But nobody could help him much in this lonely climb to the top of what seemed a wild and dangerous mountain. He would have to conquer it alone.

During the summer of 1961 things began to crystallize. Moser reached the peak. It became clear that Kolmogorov's theorem was indeed correct. The proof now seemed complete and transparent. Exploiting an idea that had suddenly occurred to him, Moser had found a method of rapid convergence. But he did not celebrate immediately. He had first to assure himself that it all fitted together correctly, for mistakes occur where one least expects them. He wanted to check everything several times. He had to wait, to put the proof aside for a while, and then return to it with fresh eyes and a clear mind.

He spent the next six months writing up his work, carefully polishing his presentation. He did not hurry. He wanted to explain everything clearly, to make things easy for the reader, to allow nonspecialists access to the proof. It was not easy, but he took his time and did it well.

People like Moser are rare. Anyone understanding the significance of such a result would have sent the paper for publication to a premier journal, to the *Annals of Mathematics*, to *Inventiones*, or to *Acta Mathematica*, seeking immediate international recognition. Anyone but Jürgen Moser. In February 1962 he submitted an article, modestly entitled *On Invariant Curves of Area-Preserving Mappings of an Annulus*, to the *Nachrichten der Akademie der Wissenschaften in Göttingen*, an obscure publication that also contains papers in the other physical and natural sciences. In fact, the paper following Moser's is a botanical study of a forest near Göttingen.

But diamonds are bright wherever the light shines. Several influential people knew of the paper, foremost among them Siegel himself, who had communicated it to the Academy in Göttingen for publication. Specialists who read it quickly appreciated its value, and shortly afterwards Moser became widely known. He was thirty-four years old.

In that same year, 1962, he received a substantial Sloan fellowship and decided to use it to finance a trip to the Soviet Union. He had met Kolmogorov and some of his students earlier that year at the International Congress in Stockholm, and he wanted to become better acquainted with them and their work. The Russians received him hospitably and Moser had fruitful discussions with them. In Moscow he spoke privately with Kolmogorov and others about the invariant tori theorem, although he gave a formal presentation on another topic. On this occasion he again met one of Kolmogorov's students, Vladimir Igorevich Arnold, who translated Moser's lecture as he spoke. With his modest knowledge of Russian, Moser observed with surprise that, during the presentation, Arnold was sometimes ahead of him with the translation.

A brilliant student of Kolmogorov, Arnold was himself interested in the invariant tori theorem. Independently of Moser and at approximately the same time, he solved the problem in a different way and under slightly different hypotheses. We will return later to a description of Arnold's work.

An acute observer of the social and political environment, Moser was pleasantly surprised to see the solid, positive development of the Soviet mathematical school, and especially the way in which talented youngsters were encouraged to develop their skills. But he also saw the darker face of the Communist regime. S. P. Novikov, today a famous topologist but in his mid-twenties and unknown at that time, asked Moser to take a paper to the West and send it to colleagues at Princeton. Novikov's work had remained unappreciated in the Soviet Union and he was struggling for recognition. His only hope was to publish in a reputable Western journal. For a Soviet scientist this was impossible without the permission of the Communist Party. One could wait years for a response to such an application, and then it would usually be negative. Moser risked provoking the Soviet authorities and fulfilled Novikov's wish. The paper ultimately appeared in the *Annals of Mathematics*. In 1970 Novikov became one of the Fields medalists announced at the International Congress of Mathematicians in Nice.

Moser's career developed rapidly. The atmosphere at the Courant Institute was ideal, and he remained there for eighteen more years, serving from 1967 to 1970 as its director. During this period he made important contributions to celestial mechanics, spectral theory, the calculus of variations, partial differential equations, differential geometry, and complex analysis. In 1980 he received an offer from the famous Swiss technological institute Eidgenössische Technische Hochschule (ETH) in Zürich. Nostalgia for his earlier life in Europe led him to accept a chair, and ultimately the directorship of the Mathematical Institute there.

Numerous prizes, medals, and honors have been awarded to Jürgen Moser, and the list is surely not complete. In 1967 he was awarded the prize of the American Mathematical Society and the Society for Industrial and Applied Mathematics, established to honor George Birkhoff. Between 1983 and 1986 he was the president of the International Mathematical Union, and he presided over the Fields Medal committee at the 1986 congress in Berkeley. In January 1989, on the occasion of his *Doctor Honoris*

Causa award from the Ruhr University in Bochum (simultaneously with two other German mathematicians, Grauert and Remmert), Moser's honorary lecture treated the subject of stability in celestial mechanics. A question he presented as of primary importance in today's mathematical research was the existence of noncollision singularities of the *n*-body problem. (The status of Xia's proof, which we discussed in chapter 3, was still not clear at that time.) Most recently, the 1994–95 Wolf Prize in Mathematics has been awarded to him.

TWIST MAPS*

The central result of Moser's 1962 paper is the *twist theorem*. We will now describe this result and, in doing so, explain the reason for its name. For this we must return to the picture of figure 5.5, showing the foliation of a three-dimensional constant energy manifold by invariant tori.

Recall that the Poincaré map was obtained by taking a cross section to the flow and by marking the points of first return to the section, as in figures 1.11 and 1.13 of chapter 1. In the present case, this means that, in the energy manifold, we take a plane that cuts the invariant tori perpendicularly. Each torus meets the plane in a circle; the section through the nested tori therefore contains infinitely many circles having a common center. (They may not be exact geometrical circles, but, as we have seen, to a topologist any simple closed curve in the plane is a circle.) Each torus corresponds to a circle, and conversely.

For the unperturbed integrable system, every orbit of the flow is confined to a torus. This means that an orbit strikes the cross section on the circle that corresponds to the torus. Continuing the journey along the orbit— iterating the first return map—we next reach the surface of section at a different point of the same circle. Thus, the points on each circle are moved around it as the first return map is iterated.

Thus, instead of studying the flow on invariant tori, one may reduce the dimension of the problem by one and consider maps that simply rotate concentric circles around their center, each circle being rotated through a different angle. Each circle is invariant: points starting on it, stay on it evermore. Choosing an exterior and an interior circle as the boundaries, Moser considered an annulus, or ring, like that in figure 5.6, and asked what happens to the circles inside this annulus when a perturbation is applied.

The problem can be further reduced to looking at a class of so-called *twist maps*, which rotate circles by an angle that increases with increasing radius. Thus, a small circle is rotated less than a large one. This derives from the assumption that the frequency ratio is nondegenerate and changes

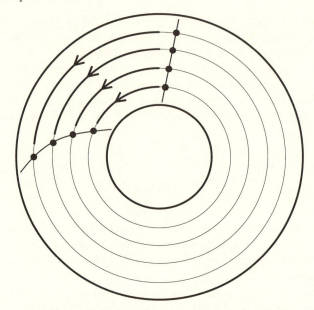

Figure 5.6. The twist map defined in the annulus.

as the value of the second integral K increases and one moves across the tori of figure 5.5. It creates a twist of the annulus, as the terminology suggests. Points on the circles corresponding to tori with rational frequency ratios are rotated through rational angles, and we speak of rational or *resonant* circles. Similarly, irrational or *nonresonant* circles correspond to those tori carrying orbits having irrational frequency ratios. The original question of perturbation of the integrable system now becomes: *If a twist map is slightly changed, will the circles still be preserved or not? Can points originally on one circle jump to another circle or not?*

We recall that Poincaré's geometric theorem, which Birkhoff proved as we described at the beginning of chapter 2, also treated mappings defined on an annulus. However, Poincaré and Birkhoff were interested in the preservation of isolated fixed and periodic points, whereas Moser was concerned with the survival of entire invariant circles: whole tori filled with orbits.

Translated into terms of the flow on invariant tori, the answer to the problem above provides a partial elucidation of Kolmogorov's theorem. Moser showed that, under certain conditions, including the requirements that the twist map be area preserving, that the perturbation be small enough and also area preserving, and that the torus is "sufficiently irrational," then at least one nonresonant invariant circle is preserved. (As we noted in chapter 2, area preservation implies that the map moves any region of the annulus to another region of the same area, as did those considered by Poincaré and Birkhoff.) The circle will possibly be deformed a little, but points on it cannot jump to other circles as the map is iterated. In fact, Moser's result implies more: that *all* nonresonant circles corresponding to those tori with frequency ratios that are sufficiently irrational are preserved. We will have more to say about this in a moment.

Though slightly different from Kolmogorov's initial statement, Moser's conclusion could be reached by imposing somewhat weaker hypotheses on the perturbative function, i.e., by relaxing the circumstances under which the conclusion is drawn. It is therefore a distinct result and it had the advantage of having a clear, rigorous proof.

Moser's crucial contribution to this problem was to construct an iterative process that converges very rapidly. Fast convergence is essential in overcoming the influence of the so-called *small divisors* or *small denominators*. These divisors had created all the troubles for almost two centuries, right from the beginning of perturbation theory. Weierstrass, Poincaré, and many others had struggled with them, but Moser was the first to really overcome the difficulties they cause.

Just as in ordinary division in arithmetic, divisors are elements that divide certain coefficients. If they are arbitrarily small, depending on the size of the coefficient to be divided, the resulting expressions may become very large. Dividing 1 by .0001 yields 10,000. In this case, the divisor is .0001; but if it is smaller, the result is much bigger. In proving that an infinite series converges, one has to show that the successive terms get smaller and smaller at a rate sufficient for their sum to be finite. If such large numbers appear as the terms of a series and one cannot control their size, the series is unlikely to converge.

Unfortunately, the perturbation methods in which one expresses quasi-periodic solutions as infinite series inevitably involve small divisors. The motions of Jupiter and Saturn around the sun provide an example. Jupiter describes 299 seconds of arc per day, while Saturn moves through 120.5 seconds. If we denote the frequencies of the two planets by ω_{Jup} and ω_{Sat} (numbers proportional to 299 and 120.5), then $2\omega_{Jup} - 5\omega_{Sat}$ $(= -4.5)$ is very small when compared with each of the planets' individual frequencies. Expressions such as the latter appear as denominators in the series for the quasi-periodic solutions of the twist map, and therefore give rise to small divisors. Laplace and Lagrange had encountered small denominators, and many pages of the former's *Mécanique céleste* contain such computations, including those for the Jupiter-Saturn resonance.

It may seem hopeless to expect convergence. However, provided the perturbing functions are smooth enough, the *numerators* of terms in the same series also decrease in magnitude as one moves to higher and higher order. It is the size of the terms as a whole, and hence the *relative* rates of decrease in their numerators and denominators, that determines convergence. Here the notion of a sufficiently irrational frequency ratio enters. As we have noted, arbitrarily small denominators are unavoidable, for arbitrarily close to each irrational circle there is a rational one. However, if the irrational ratio ω_1/ω_2 can be approximated closely only by fractions m/n with very large integers m and n, it turns out that the decrease in numerators beats the decrease in denominators, so that the series converges, implying that the corresponding circle survives.

In taming the small divisor problem, Moser answered a long-standing question that had been tackled without success by many of his predecessors. Big ideas are almost always direct and simple, and Moser's insight had both these qualities. His work did not stop here. He continued to attack difficult questions, in other fields of mathematics, with the same flair. Those who follow Moser's achievements admire the natural way in which the ideas spring from the context. Reading one of his papers is like climbing a dangerous mountain on the safest and most picturesque path. Once on top, your efforts will be rewarded with magnificent views. And the one who discovered and marked the trails is Jürgen Moser.

A GIFTED STUDENT

Andrei Nikolaevich Kolmogorov raised his eyes from the manuscript and looked at his student. "Young man," he said carefully, "you have solved Hilbert's thirteenth problem!" Vladimir Igorevich Arnold's heart began to beat faster than usual.

In his plenary lecture at the International Congress of Mathematicians held in 1900 in Paris, the great German mathematician David Hilbert had outlined a number of themes and specific areas that he judged particularly important for the future development of mathematical research. He composed a list of twenty-three specific questions: conjectures and problems that captured the essence of the subject. Hilbert is perhaps best known in the history of mathematics for his espousal of the *formalist* approach: the belief that all the statements of number theory, and ultimately analysis as a whole, could be shown to be internally consistent, a hope that was finally dashed in 1931 by Kurt Gödel's famous Incompleteness Theorem. Nonetheless, Hilbert's list of problems has been very influential in determining the direction of research in mathematics—some might say too influential. Though a vast amount of energy has been spent in attempts to solve them, several remain unsolved today.

In 1956, in a moment of inspiration, a Russian teenager had achieved what many others had attempted unsuccessfully for over half a century.

Plate 5.4. Vladimir Igorevich Arnold. (Courtesy of S. Zdravkovska)

Vladimir Igorevich Arnold was barely nineteen years old when he showed his teacher the notes of his solution of Hilbert's problem. Hoping to stimulate his students to further research, Kolmogorov had presented in one of his seminars the "reduction"—a restatement in somewhat simpler terms—that he had found for the thirteenth problem. This reduction had itself required many months of work, so he little expected that one of them would appear one day with a solution. The problem itself falls outside the scope of our story, and we shall not describe it, except to note that Arnold had succeeded where his elders had failed, and at a time when youths of his age are normally just finishing high school.

From that time on, it was clear that Arnold would become a mathematician. He had the good fortune to have been born at the right time, when this path was open for him. Having a Jewish mother, he would not have been permitted to study at the University of Moscow during the Stalinist period. The death of the Soviet leader in 1953, followed by the political relaxation that lasted until the invasion of Czechoslovakia in 1968, gave him just enough time to obtain his degrees and become a professor in the Department of Mathematics and Mechanics. He was indeed fortunate to reach the appropriate age during the only open period known by the Soviet Union in its over seventy years. Arnold knew how best to use those precious years.

After the success with Hilbert's problem, Kolmogorov encouraged his student to choose his own research topic. Kolmogorov's approach to teaching was liberal; he gave his students the freedom to work on whatever they wished, requiring only that they inform him when they obtained something interesting. (He did, however, have definite tastes: when Arnold subsequently studied cardiac oscillations, Kolmogorov discouraged him, saying: "That is not one of the classical problems one ought to work on." Mathematical physics was much more important!) In class he spent little time on formal proofs, presenting only the essential features: the kernel of the problem. Students were supposed to fill in the gaps themselves, at home. Kolmogorov took the same approach in his own research. He seldom bothered to write down all the details. He felt life was too short and preferred to spend it creating, not grubbing. Few mathematicians share this point of view, since it encourages superficiality. Moreover, it can lead to serious confusion, as Moser discovered in reading the "proof" of the tori result.

It would certainly be dangerous if most mathematicians proceeded in this way, for then one could not readily distinguish between proven theorems, unproven (but true) statements, false claims, and conjectures. The vast majority of published mathematical papers are correct and completely reliable, in contrast to other scientific fields, in which the status of new discoveries is sometimes unclear. This enviable state of affairs relies on "normal" mathematicians painstakingly working out and checking every detail. But in exceptional cases a more cavalier approach has accompanied unusual creativity. It did not prevent Kolmogorov from becoming successful, and he was by no means superficial. In fact, Yasha Sinai, who as an undergraduate attended Kolmogorov's course and seminar in 1956–1957 along with Arnold, remarks that it was clear to the participants that Kolmogorov knew how to fill in all the gaps in his proof of the tori result, with possibly one exception, and this he was convinced was possible too. It concerned a technical detail regarding the measure of the surviving tori. Indeed, Moser and, as we shall see, Arnold *were* both able to provide the missing details for this step and to arrive at independent proofs of the whole.

Following Kolmogorov's encouragement, Arnold did not wait for further advice. He immediately began to work on another problem related to the one he had just solved. It concerned whether *a function on a curve can be represented as the sum of functions of its coordinates.* To explain this, suppose we have a function *f* defined on the plane that also takes values in the plane (both the input and output are two-dimensional). So f has two output coordinates: f_1 and f_2. If we fix a curve in the plane and *restrict f* to this curve (i.e., consider the values of the function only on this curve, ignoring those elsewhere), under what circumstances can we write f_1 and f_2 as the sums of functions of the planar input coordinates x_1 and x_2? Arnold first considered the simpler case in which the fixed curve is a circle. He quickly realized that the question implies the study of iterations of maps on the circle. Then he saw that this reduces further to exactly the problem of the invariant tori. This was indeed surprising, since he had started from a domain apparently unrelated to the theory of dynamical systems and managed to find a bridge between the two fields.

Arnold then went on to tackle the invariant tori problem itself, and in 1961 he produced complete proofs of Kolmogorov's statements under hypotheses different from those imposed by Moser. (Arnold assumed analytic perturbations; Moser, sufficiently smooth ones.) Some of these results were presented at conferences of mathematicians and astronomers in Moscow in the summer and fall of 1961, and papers concerning "model problems" also appeared that year. The first full announcement of the result for systems having many degrees of freedom was published in *Soviet Mathematics Doklady* the following year, coinciding with Moser's *Nachrichten* paper. Arnold's title, "The Classical Theory of Perturbations and the Problem of Stability of the Solar System," clearly reveals the paper's place in the three-century history that we have traced in this book. In this respect it is ironic that Arnold has also remarked that "the 200-year interval from Huygens and Newton to Riemann and Poincaré seems to me to be a mathematical desert filled only with calculations."

Further, more extensive papers in the English translation *Russian Mathematical Surveys* of 1963 presented the details and established Arnold as a co-creator of KAM theory. Note that we do not speak of a "KAM theorem," since there is rather a body of theory containing several distinct results, of which Kolmogorov's, Arnold's, and Moser's are the earliest. Many technical refinements and generalizations have now appeared, and KAM theory remains a very active area of research.

CHAOS DIFFUSES

A second major contribution to classical mechanics with perhaps even more striking and unexpected physical implications came in 1964, when Arnold discovered the fascinating phenomenon that now bears his name: *Arnold diffusion*. Until he constructed the first (and somewhat artificial) example, his adviser Kolmogorov had strongly believed that such behavior is unlikely to occur in Hamiltonian systems. As we shall see, this property is complementary to the stability results of the KAM theory. It implies the presence of a particular form of chaos, with highly unstable orbits coexisting among the stable quasi-periodic ones, so it is small wonder that Kolmogorov was reluctant to accept its existence before Arnold showed that it could indeed occur.

Arnold diffusion concerns the possibly chaotic behavior of orbits that originally lay on resonant tori and on certain nonresonant ones: the tori that are destroyed when the small perturbation is applied. In the example pictured in figure 5.5, these tori lie between "stable" nonresonant ones, which are *not* destroyed by small perturbations. Thus, orbits originally lying on resonant tori cannot escape from the region between two neighboring stable ones: they remain trapped between those two surfaces. This two-degree-of-freedom example therefore suggested that *all* orbits would manifest some kind of stability. Kolmogorov believed this should also hold for higher-dimensional tori in systems with three or more degrees of freedom. Arnold disagreed.

Before describing Arnold diffusion, it is worth exploring further the two-degree-of-freedom case of figure 5.5. We shall describe its behavior in terms of the Poincaré mapping on the annulus produced by taking a cross section in a constant energy manifold. The structure of the unperturbed map of the integrable system was illustrated in figure 5.6. As we saw, it is a simple twist, every circle being invariant, and its points rotated through a greater or lesser angle by the map, depending upon its radius. We recall that the KAM theory implies that, after perturbation, most of the circles having irrational frequency ratios are preserved as invariant curves, albeit in mildly distorted form. The intervening rational circles, and some of the irrational ones, are destroyed.

Typically the rational circles break into discrete sets of periodic points, some of which are stable and others of saddle type. Again typically, the stable and unstable manifolds of the latter intersect transversely, so that the resulting homoclinic points create precisely the kind of chaos we explored in chapters 1 and 2. Nonetheless, each such set of periodic points, with its intersecting manifolds, homoclinic tangles, and chaotic orbits, is trapped between a pair of neighboring invariant circles that survive, and so we have a kind of coexistence of order and chaos. We attempt to illustrate the structure in one such *resonant band* in figure 5.7. This is presumably the picture that Poincaré thought too complex to draw, and, as in so much else, he was correct. In our sketch we show only the first stage of an infinite sequence of periodic points and stable and unstable manifolds, all of which are packed into the homoclinic tangles between each pair of surviving invariant circles. And as we have seen, an infinite Cantor set of circles is preserved.

Thus, the reader should try to imagine infinitely many such bands, corre-

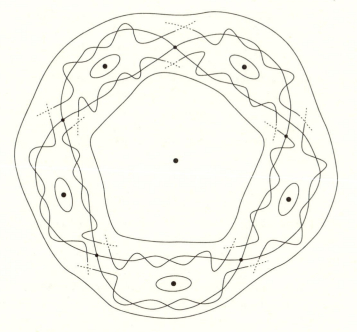

Figure 5.7. Coexistence of stability and chaos in a perturbed annulus map.

sponding to the gaps in a Cantor set like that of figure 3.7, with the circles separating them as the points in the Cantor set. Chaotic and stable motions are intimately interleaved, as we promised at the end of chapter 4. The former are chaotic in the sense that they circulate irregularly within each band, hesitating for greater or lesser periods near unstable saddle points, but they are orderly in the sense that they cannot escape from the band within which they start. Viewed from a distance, it seems a mild kind of chaos. On this basis, we can now describe Arnold diffusion, which is significantly stranger and stronger.

As we have seen, for a conservative system the total energy remains constant on orbits, and so solutions are confined to an energy manifold of one dimension lower than that of the full phase space. A three-degree-of-freedom system, with three position and three momentum variables, therefore has a family of five-dimensional energy manifolds. Each of these contains three-dimensional tori, there being generally three independent frequencies. It is natural to ask if these tori provide barriers in the energy manifold that solutions cannot cross, as the two-dimensional tori clearly do in the three-dimensional case of figure 5.5.

We cannot draw pictures in spaces having dimensions greater than three, so we will use an analogy involving lower-dimensional objects. Suppose that, before perturbation, the resonant and the nonresonant tori were concentric circles lying in a plane, and suppose that the phase space were three-dimensional as in figure 5.8. After perturbation, the resonant and some of the nonresonant circles break up, while most of the nonresonant ones deform slightly but remain closed curves. The difference between this example and the one in figure 5.7 is that, in this case, the "broken orbits" are no longer necessarily trapped between closed curves, for now they can leave the plane and move freely in space. The surviving circles, being one-dimensional, cannot separate different regions of three-dimensional space.

Actually, in his original paper submitted for the prize described in chapter 1, Poincaré had already realized that the existence of smooth invariant surfaces in systems having more than two degrees of freedom would not necessarily imply stability. There is certainly enough "room" in these higher-dimensional phase spaces for orbits to sneak past the tori, just as we can step outside a circle painted on the pavement. The question becomes: do any orbits of the actual differential equation take advantage of this freedom to leave the region in which they started, and if so, how do they behave after their escape?

Arnold constructed an example that answered the first of these questions in the affirmative. Moreover, he showed that the behavior of the escaping

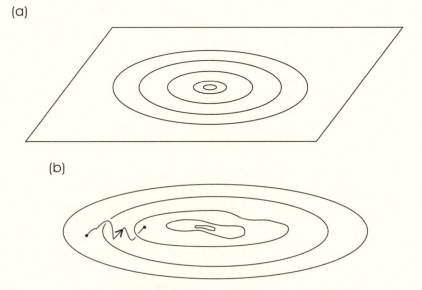

Figure 5.8. An example suggesting the behavior of tori in higher dimensions: (a) before perturbation; (b) after perturbation, showing an orbit escaping.

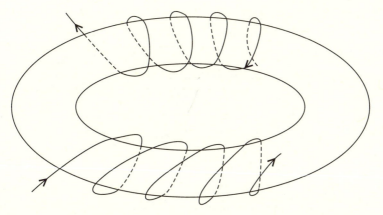

Figure 5.9. An orbit belonging to the stable manifold and an orbit belonging to the unstable manifold of the torus.

orbits is "very chaotic" in a sense we will now try to make clear. His example involved three degrees of freedom, and he considered two-dimensional tori with three-dimensional "whiskers." He generalized the two-degree-of-freedom case of figure 5.5 in the following sense. In that case, the resonant two-dimensional tori break to form one-dimensional periodic orbits (circles), which appear in the Poincaré map of figure 5.7 as the periodic points having stable and unstable manifolds. For three degrees of freedom, the resonant three-dimensional tori partially break to form two-dimensional tori.

Just as saddle point equilibria of a differential equation or map can possess stable and unstable manifolds filled with orbits that approach and leave the equilibrium, as in figures 1.6, 1.9, and 5.7, so can tori of quasiperiodic orbits. There may exist orbits tending toward a torus, as in figure 5.9, to reach it after an infinite length of time. The set of all orbits with this property forms a (hyper-) surface that, if certain conditions are fulfilled, is a smooth, stable manifold. Similarly, orbits leaving the torus belong to its unstable manifold. In Arnold's example, the tori were two-dimensional, as in figure 5.9, and their stable and unstable manifolds, or *whiskers*—as he more suggestively named them—were each three-dimensional. Our picture takes a liberty here, since we cannot really show orbits in both the stable and unstable whiskers simultaneously in the same three-dimensional space. Again, the "size" of high-dimensional phase space is necessary to fit everything in.

The stable and unstable manifolds have dimensions that can vary from torus to torus. Looking at the lower part of figure 5.9, one can imagine other orbits tending toward the torus, filling out a full three-dimensional neighborhood. In fact, the stable and unstable manifolds of a k-dimensional in-part of the stable manifold

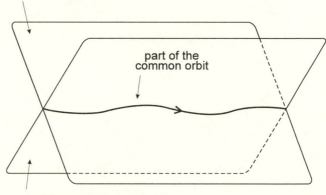

part of the common orbit

part of the unstable manifold

Figure 5.10. Transversal intersection of the stable and unstable manifolds of a torus along a common orbit.

variant torus in an n-degree of freedom system might have any dimension up to $n - k$. The stable and unstable manifolds can intersect each other along an orbit, and this intersection may be *transversal*, as in figure 5.10, in which each of the manifolds is drawn as a surface. The meaning of "transversal" is the same as in chapter 1, i.e., that the two surfaces cross each other nontangentially along a common orbit. Using similar reasoning to that in chapters 1 and 2, we can conclude that transversal intersection of the stable and unstable manifolds of an invariant torus leads to chaotic behavior similar to that associated with orbits homoclinic to a fixed point.

Arnold diffusion is richer still. It occurs when *multiple* transversal intersections among the stable and unstable whiskers of a finite sequence of *distinct* tori occur, so that one finds orbits that skip from torus to torus more or less "at random." Arnold called this a *transition chain*. To picture it, we will represent the tori by points and their stable and unstable manifolds by curves. (Of course, the dimensions of the tori and of the manifolds can be high).

In figure 5.11 we show how the transversal intersections cause Arnold diffusion. The stable and unstable manifolds of the tori T_1, T_2, and T_3 intersect each other as indicated. An orbit starting near T_1 can follow the heteroclinic orbit almost to T_2, where it transfers its allegiance to the next orbit, leading to T_3. This may continue indefinitely. The picture is "filled in" under iteration of the map, as in figure 5.12, which may suggest the complications of the chaotic behavior. At each encounter with a torus, the chaotic orbit may go on to the next torus in the chain, or it may return to revisit an earlier one. However, in contrast to the faithful representations of the three-

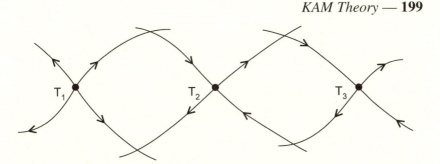

Figure 5.11. Transversal intersection of stable and unstable manifolds for a finite sequence of tori gives rise to Arnold diffusion.

dimensional phase space of the pendulum and its Poincaré map in figures 1.12 to 1.14, figures 5.11 and 5.12 are only schematic. Arnold diffusion requires at least three degrees of freedom: a five-dimensional energy manifold with a four-dimensional Poincaré map.

Arnold's example of 1964 was artificial in the sense that the differential equation he constructed to show the existence of diffusion has no obvious physical significance. In a footnote to the same paper, however, he conjectured that this kind of chaos would also occur in the three-body problem. Unfortunately, it is very difficult to prove that such a complicated behavior actually does take place in specific models of physical systems. There has been some progress in the last thirty years, but only for rather special cases. Certain mechanical systems of coupled oscillators, for example, are now known to exhibit diffusion. Here it implies that the energy can pass back and forth from oscillator to oscillator (mode to mode) as if at random. Although it is believed that Arnold diffusion is a generic property for nonintegrable Hamiltonian systems, no one has so far obtained conclusive results. Also, most of the evidence suggests that, when Arnold diffusion does occur, it takes place over enormously long timescales. It is therefore a subtle effect, difficult to recognize and quantify in physical systems.

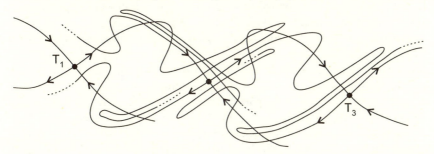

Figure 5.12. A further approximation of the chaos produced by the Arnold diffusion phenomenon.

However, Zhihong Xia, the same who earlier solved Painlevé's conjecture, has recently demonstrated that Arnold diffusion is present in the three-body problem. He proved this result in stages, by first dealing with a special case of the restricted three-body problem and then extending his ideas to the general case. This represents another noteworthy contribution of the young Chinese mathematician to a problem that had remained unsolved for almost three decades.

Perhaps more strikingly, Luigi Chierchia and Giovanni Gallavotti of the University of Rome have recently shown that Arnold diffusion occurs in a model for the rotation of a slightly asymmetrical oblate planet in orbit around a fixed star—a problem originally due to d'Alembert. Their 144-page paper, a triumph of classical perturbation theory and hard analysis, was awarded one of the 1995 prizes endowed by the Institut Henri Poincaré and by Gauthier-Villars (Poincaré's publisher)—an especially fitting tribute.

There is some evidence that Arnold diffusion is implicated in another classical problem in the astronomy of the solar system. In 1857, Daniel Kirkwood, an American mathematics and astronomy teacher, observed several gaps in the belt of asteroids between Jupiter and Mars. In this respect the configuration of the asteroid belt is similar to that of Saturn's rings, although in the latter the particles are much smaller and closer together. Since then, astronomers and celestial mechanicians have devoted much energy to attempts to understand how these gaps developed and whether they are long- or merely short-term features. The existence of some gaps is explained by a so-called *resonance phenomenon*, which stems from the small denominator problem discussed earlier in this chapter: a large perturbing force occurs at just that distance from the sun at which frequencies are perfectly matched, thus throwing the particles out of their orbits in that region. Unfortunately, there is no general explanation for all of the Kirkwood gaps, and most of the existing results are based on numerical simulation rather than rigorous mathematical proofs. Interested readers can learn more from Ivars Peterson's book, cited in the bibliography.

Xia's recent theorem on the three-body problem holds out hope to those suspecting that Arnold diffusion might explain the Kirkwood gaps. A three-body problem involving Jupiter, the sun, and an asteroid might reveal how, for certain initial conditions, the asteroid leaves the belt to which it originally belonged. Such an explanation would be consistent with earlier results. Unfortunately, the task is not that simple. Xia's qualitative result does not apply directly to the particular quantitative problem of Jupiter and the sun. In fact, his proof, when applied to the solar system, shows that Arnold diffusion occurs for particles that would originally orbit in a region *beyond* Jupiter. Of course, this does not mean that Arnold diffusion cannot occur inside the orbit of Jupiter: it merely implies that Xia's proof is not directly applicable to the asteroid belt.

Epilogue

The world of celestial mechanics is still thriving more than three hundred years after it began with Isaac Newton's *Principia Mathematica*. The reader who has followed us this far will no longer think it strange that a single mathematical problem—the solution of the differential equations describing the *n*-body gravitational problem—should have occupied the minds of so many illustrious figures in the history of mathematics, or that it should continue to fascinate and frustrate us today. As we have seen, to solve the equations of dynamics is not one problem, but many. It is perhaps not so much these problems themselves, as all their unexpected by-products, that have had such an influence on the development of mathematics and science in general. Good mathematics, like all good science, is often largely concerned with making connections between ostensibly disparate pieces of information and fields of knowledge, and we have seen again and again how this has occurred in the mathematical theory of celestial mechanics.

Dynamical systems theory, which Poincaré essentially invented in his (revised) prize memoir as he struggled to understand homoclinic orbits, has grown today into the vast, disorganized, but vital subject of *nonlinear dynamics*, a field that touches almost all areas of engineering and the sciences in which differential equations are used. It is a collection of rigorous results and tools such as symbolic dynamics, of computational and geometric methods, and of many looser ideas and conjectures. Chemists trying to understand the patterns formed on surfaces of catalysts, engineers studying the instabilities of railroad cars, biologists and physicians treating heart disease, physicists observing crystal growth, aircraft designers concerned with turbulence, meteorologists, and climate modelers: all are finding applications for the insights that had their origins in the "simple" *n*-body problem. Books such as those of Leon Glass and Michael Mackey, James Gleick, Edward Lorenz, David Ruelle, and Ian Stewart, listed in the bibliography, will tell the reader more about these things.

But dynamical systems theory has had perhaps an equally important influence within mathematics itself. We have already remarked that Poincaré was also a founder of topology, which, along with analysis and algebra, is one of the central pillars of mathematics. This was no coincidence. One can characterize toplogy as the study of continuous mappings relating objects in (topological) spaces X and Y, as $f: X \rightarrow Y$. When the spaces are the same, we can repeat the application of f, iterating the map, and we have a dynamical system. Stephen Smale, whose invention of the horseshoe map we described in chapter 2, made contributions to topology at least equal to those in dynamical systems, and many of his ideas relied on taking a dynamical viewpoint. More recently he has become interested in computation and issues of computability and complexity, and again a dynamical perspective has proved useful. Most scientific computation, after all, involves repetitive cycling to converge on an answer. Computer centers even speak of selling "cycles" on their machines. The proceedings of the conference held in 1990 in Berkeley to celebrate Smale's sixtieth birthday are called *From Topology to Computation*, and it was argued by several people at the meeting that dynamics is the unifying thread in his work. Smale is unusual in the depth and breadth of his interests, but a number of other mathematicians have found that dynamical systems theory points toward a more unified view of their subject.

In our haste to press on with new applications and insights, let us not forget from where we came. As we celebrate the sudden outpouring of ideas and the ferment of nonlinear dynamics, we must remember its origins, for in them we find both strengths and limitations. Before plunging into nonlinear studies or going hunting for chaos in a favorite problem, we should reflect on the fact that ideas and properties that appear to have purely physical bases, such as stability or even chaos itself, demand precise mathematical definitions if they are to be usefully applied. It is of little value to say that an airplane is stable if we have not agreed on a common meaning for the notion of stability, and our claim becomes much more convincing if we prove that a mathematical model of the airplane has solutions that are resistant to sudden perturbations.

The bulk of the rigorous mathematical foundations of the field of dynamics has its origins in the people and works we have described here. We hope that this book has gone some way toward revealing and explaining these vital roots.

Notes

THE FOLLOWING notes provide bibliographical and historical details. Recognizing that many readers are uninterested in such detail, we have appended them after the main text—rather than as footnotes—so as not to interrupt the story. They are intended primarily to help the mathematician or historian of mathematics locate references.

This book was conceived as a work of popularization and we believe there is an interesting story to tell, so for continuity's sake we have sketched in some scenes without full documentary support. The few cases in which we permitted ourselves such liberties are clearly pointed out below. Our excuse is that we wish to introduce the reader to the atmosphere and the spirit of the times when the events in question took place, and in doing so provide a more vital background for the facts.

The notes are organized in sequence with the sections of each chapter, referenced by two numbers (for example, 2.3 means the third section of the second chapter). We also give the title of the corresponding section. Bibliographic references (quoted in brackets) follow the notes; some of these books and articles contain general background and not all are explicitly referred to. The authors' names—abbreviated as FD and PH—occur when we draw on discussions and other unpublished sources; in the interests of precision, we describe the circumstances in these cases. Since some of the materials we have used are parts of our general knowledge, and the events surrounding their gathering may be blurred in our memories, we would like to apologize to anyone we have inadvertently misquoted or neglected to mention.

The first epigraph to the book as a whole is translated from lines 114–116 of the Theogony of Hesiod. It was suggested to us by Stathis Tompaidis. The second is from [Dy,1995].

Chapter 1.
A Great Discovery

1.0. The epigraph to chapter 1 is taken from [Da,1900]. It is also cited in the introduction to [Po,1929]. In notes on a seminar on *Les methodes nouvelles* run by Jacques Laskar and Alain Chenciner at the Bureau des Longitudes in Paris in 1989, the same remark (in French) is credited to Paul Appell in 1925.

1.1. A WALK IN PARIS: The first part of the chapter is built around Poincaré's struggle to understand the problem of homoclinic points and chaos, and the sudden comprehension that came to him while walking and thinking of something else. There is no historical evidence for this. Nevertheless, Poincaré describes a similar event connected with the discovery of a property related to Fuchsian functions, when, away from his desk, relaxed and detached from the problem, the idea of the proof struck him from nowhere ([Po,1929], pp. 387–388). This is by no means unusual for a scientist—Kekulé's discovery of the benzene ring is a similar instance. We therefore thought that it would be an appropriate way to introduce the story. It is certainly true that Poincaré did not realize the implications of his discovery of homoclinic points in 1889 until some years later, and that they caused him worries for many years.

Poincaré indeed had a happy marriage, which produced three daughters and a son [Be,1937]. In July 1993, at Oberwolfach, Germany, Marc Chaperon of the University Paris VII told FD that a descendant of Henri Poincaré had been one of Chaperon's undergraduate students. Other references on Poincaré used in this and the opening section are [Nm,1960], [Be,1937],[*] and [Co,1990]. The facts concerning Gauss, Bolyai, and Galois are traditional in the history of mathematics, cf. [Nm,1960].

1.2. Newton's Insight: The major source is [N,1934], the original publication being [N,1686]. For information on Newton's life and achievements, including his dispute with Leibnitz, see [We,1980]. The information about Napier and the first differential equations was provided to FD by the historian of mathematics Gary Tee in January 1993, at a conference on differential equations and scientific computation held at the University of Auckland, New Zealand.

1.3. A Language for the Laws of Nature: The ideas described in this section can be found in most modern textbooks on differential equations. See [Ar,1973] or [H & S,1974], for example.

1.4. Models of Reality: The same references as in 1.3.

1.5. Manifold Worlds: The concepts of this section are to be found in introductory textbooks on topology or modern differential geometry as well as in books treating differential equations on manifolds (see 1.3).

1.6. The N-Body Problem: The n-body problem was formulated for the first time (in geometrical terms) by Newton [N,1686,1934]. For the statement that the two-body problem was solved by Bernoulli, see the bibliographical notes in [Win,1941].

1.7. King Oscar's Prize: Goroff's introduction to [Po,1993] is an excellent source on the history of King Oscar's prize; see also [S & M,1971] and the historical articles [B-G,1994] and [And,1994]. The competition was announced, in French and German, in *Acta Mathematica*, vol. 7, of 1885–86. The problem posed by the jury is cited by Gaston Darboux in [D,1914] and was translated into English in the form cited here in [Po,1993].

1.8. Poincaré's Achievement: The words of Poincaré, "Life is only a short episode . . . ," appear in [Po,1929]. For the early papers on celestial mechanics, see [Po,1951-6]. The translation of Poincaré's epigraph appears in [B-G,1994].

---

[*] Although Bell's book is criticized by some historians as often being overly imaginative, we found it useful for our purposes. A recent review [Br,1994] characterizes Bell's work as belonging to a category of scientific books that describe the "mathematical truths that do not fit into the existing matrix of tradition and expectation," concluding that "we need mathematicians like Bell . . . who can draw on the poetic muse."

1.9. LES MÉTHODES NOUVELLES . . . : Poincaré's prize memoir was published as [Po,1890]. *Les méthodes nouvelles* appeared in three volumes as [Po,1892–3–9]; an English translation by D. Goroff has only recently been published [Po,1993].

1.10. FIXED POINTS: A more technical introduction to the ideas in this section and parts of chapter 2 can be found in [Ho,1990].

1.11. FIRST RETURNS: The sources are the same as 1.9. The idea of the return map was first introduced in [Po,1881–6].

1.12. A GLIMPSE OF CHAOS: See [Ho,1990] and [Po,1993], from which the description of the homoclinic tangle is quoted, in Goroff's translation.

1.13. PANDORA'S BOX: The only references here are [Po,1892–3–9] and [Po,1993]. The famous statement on the homoclinic tangle occurs in chapter 33, section 397.

1.14. POINCARÉ'S MISTAKE: For further information, the reader should consult the collection of letters from Poincaré to Mittag-Leffler, printed in *Acta Mathematica*, vol. 38; Goroff's introduction to [Po,1993]; and [Mou,1912]. An earlier account of events surrounding the mistake appeared in [Pet,1993]. Richard McGehee told PH at the AMS meeting in Phoenix, Arizona, in January 1989, of his discovery of the original printings of Poincaré's paper. Several discussions between FD and McGehee at the Dynamical Systems Conference in Oberwolfach, Germany, July 1993, also helped clarify the story. Two recent papers, [B-G,1994] and [And,1994], which benefited from access to archives in the Mittag-Leffler Institute, were invaluable in providing additional detail. Of the papers referred to in the text, Gylden's is [Gy,1887] and that of Buchholz is [Bch,1904]. We thank Jürgen Moser for drawing our attention to the latter. ([Bch,1904] also remarks in passing that the purse was 10,000 krone, not 2,500, as stated in the announcement; cf. 1.7 above, but the latter figure is accepted by [B-G,1994] and [And,1994].) Moulton's opinion [Mou,1912] on Poincaré's work is reproduced in [Po,1993].

1.15. A SURPRISING DISCOVERY: This section is based on the discussions with Richard McGehee mentioned above. A description of the library at the Mittag-Leffler Institute and a drawing representing the former residence of Mittag-Leffler can be found in [Bö,1992]. Some of Poincaré's general writings on science (most of them also published separately) appear in [Po,1929].

Chapter 2.
Symbolic Dynamics

2.0. The epigraph to chapter 2 is taken from [Mor,1946].

2.1. A FIXED POINT BEGINS A CAREER: The introductory scene is the product of our imaginations and was included to suggest the state of mind of a scientist on the verge of receiving a possible job offer. We cannot know if Birkhoff responded in this way to Harvard's offer (perhaps he was less excited than we might have been). For the rest, see [Bi,1913, 1927, 1935, 1968] and Goroff's introduction to [Po,1993]. Poincaré's last paper is [Po,1912]; it can also be found in [Po,1951–6].

2.2. ON THE BEACH AT RIO: For J. L Synge's correspondence on the Fields Medal and related information, see [Sy,1933] and [Tr,1976]. For remarks on Nobel, see [Mey,1994].

The bulk of this section is based on an article that appeared in [Sm,1980] and is reprinted in [HM & S,1993]. Smale provided PH with some further details in a telephone conversation on 3 January 1995. Also see [Sm,1991]. The papers of van der Pol & van der Mark, Cartwright & Littlewood, and Levinson are [V & V,1927], [C & L,1945], and [Levin,1949]. Dame Mary Cartwright described her wartime work and other recollections on dynamical systems theory in [Ca,1972]. The precise way in which horseshoes are embedded in the Poincaré map of the van der Pol equation was not elucidated until the late 1970's [Levi,1981].

2.3. SMALE'S HORSESHOE: The references concerning Smale are those in 2.1. The mathematical results on the horseshoe can be found today in most modern textbooks on dynamical systems, such as [Mos,1973] and [G & H,1983].

2.4. SHIFTS ON SYMBOLS: There are several books discussing symbolic dynamics for example, [De,1986] and [G & H,1983]. An early text is available in the mathematics library of the Institute for Advanced Study in Princeton: the notes by Rufus Oldenburger from lectures given by Marston Morse in the academic year 1937/38, [Mor,1938]. A major review paper of Smale's on dynamical systems, which appeared initially in the *Bulletin of the American Mathematical Society*, is reproduced in [Sm,1980], where supplementary notes and comments also appear. Smale's first paper on the horseshoe appeared in a collection of papers published by Princeton University Press entitled *Differential and Combinatorial Topology* [Sm,1965]; it is referenced in [Sm,1980]. For the properties defining chaos, see [De,1986] and [B et al.,1992].

2.5. SYMBOLS FOR CHAOS: See the references mentioned in 2.4. The early remark of Maxwell on sensitive dependence on initial conditions is cited in [Ber,1978] and [Ek,1988]; see also [H & Y,1993]. Lorenz's address, which was previously unpublished, is included in [Lo,1993].

2.6. OSCILLATIONS AND REVOLUTIONS: The papers, books, and proceedings mentioned here are [Bi,1927, 1935], [Sit,1961], [A,1968a, 1968b, 1969]. An account of Sitnikov's work appears in [Mos,1973]. The proceedings of the Yale conference were published as [Wa,1993].

2.7. A NEW SCIENCE?: For examples of the range of applications of nonlinear dynamics in the sciences, see [Gl,1987] and [St,1989]. [Hay,1990] discusses the impact of chaos on literary studies.

Chapter 3.
Collisions and Other Singularities

3.0. The epigraph to chapter 3 is taken from Garnier's introduction to [Pa,1972]. Some of the material in this chapter first appeared in [Di,1993b]. The paper [S & X,1995], written for a mathematical audience but in relatively nontechnical form, also contains a description of the key results leading to Xia's proof of noncollision singularities in 3.10.

3.1. A SINGULAR MAN: The king's attendance at Painlevé's opening lecture is described in [Pa,1972], [S & M,1971]. Further information on his support for mathematical activities appears in [B-G,1994]. Painlevé's words are translated from the published version of the same lecture in [Pa,1972], and the handwritten manuscript, eventually published as [Pa,1897], is also reproduced in [Pa,1972], vol. I. A biography of Painlevé can also be found in [Pa,1972].

3.2. COLLISION OR BLOWUP: We have no clear evidence that von Zeipel attended the opening lecture, but it is quite reasonable to suppose that his early decision to study celestial mechanics was influenced by Painlevé's presence in Stockholm. The other data on von Zeipel's life were taken from [Mc,1986]. The history of von Zeipel's noncollision result appears in the notes of [Win,1941]. Other relevant references are [Ch,1920], [Sp,1970].

3.3. COMPUTER GAMES: For an account of von Neumann's involvement in the early programmable computer built at the Institute for Advanced Study, see [Re,1987]. The information on Szebehely's team and the others is to be found in [Sz,1967]. Comments about Meissel's computational attempts were provided to FD by Jaak Peetre in February 1995, via E-mail (a scientific biography of E. Meissel appears in [Pe,1995]). The numerical computations in figure 3.4 also appeared in [Ar,1988]. A proof of the ejection effect which takes place, in general, after the occurrence of a triple-collision approach appears in [Mc,1974] and [W,1975,1976].

3.4. HOW TO CATCH A RABBIT: The papers containing the research described in this section are [P & S,1968, 1970], [S,1971, 1973a, 1973b, 1973c]. The properties proven by Donald Saari in 1972 were published in [S,1973b]. The background to the story was filled in by Saari in the spring of 1991 during a visit at Centre de Recherches Mathématiques in Montreal, where FD was a postdoctoral fellow. Additional details were obtained by FD from Saari in 1993 via E-mail, and in June 1995 at the Joint Summer Research Conference in Mathematics: Hamiltonian Systems and Celestial Mechanics, University of Washington, Seattle. A discussion between FD and Carl Simon on the subject also took place at the Midwest Dynamical Systems Conference held in October 1993, in Berkeley, in honor of Morris Hirsch.

3.5. A MEASURE OF SUCCESS: This section is also based on the sources mentioned in 3.4 and on the additional papers [S,1975, 1976, 1977, 1978, 1984]. It is worth mentioning some recent results on the gravitational model proposed by the Bulgarian physicist Manev (or Maneff in German or French spelling) in the twenties [Ma,1924, 1925, 1930a, 1930b]. These results show that, in this model, the set of initial data leading to collisions has *positive* Lebesgue measure (see [Di,1993a], [Di,1996b]). Manev's model is in a certain sense an approximation of relativity and provides good agreement with observed astronomical phenomena. Einstein used it, in a simplified form, to approximate relativity and to prove that the perihelion advance of Mercury can be computed with great accuracy. This suggests that collisions may not be so improbable in the universe. Other types of models, such as the one proposed by Mücket and Treder in 1977, seem to behave qualitatively in the same way as the classical Newtonian one; see [Ba & D,1993], [Di,1996a]. For Saari's work on voting and apportionment, see [S,1978], [S,1987], and [S,1994]. Pancoe's Master's thesis is [Pan,1951]. Littlewood's book has been reprinted, with additions [Lit,1986].

3.6. REGULARIZING COLLISIONS: Information for this section was provided to FD by Richard McGehee during the Berkeley conference mentioned in 3.4. Siegel's paper mentioned here is [Si,1941]. Sundman's paper is [Su,1912]; in fact, the research published in that article is an improved version of two earlier papers: [Su,1907] and [Su,1909]. Moser's book is [Mos,1973].

3.7. CELESTIAL BILLIARDS: This section is also based on the discussion with Richard McGehee mentioned in 3.6. Easton's paper on block regularization is [E,1971]. Levi-Civita's early transformations, which stimulated the research of

McGehee in [Mc,1974], were published in [Le,1920]. Comments on Conley were provided by Chris Jones and Catharine Anastasia Conley in E-mail messages to FD and PH in May and June 1995. The story about phase portraits comes from Marty Feinberg via Chris Jones; they and Neil Fenichel all stressed Conley's generosity and desire for others to take credit. For a computer realization of Conley index theory, see [Mi & Mr,1995].

3.8. ENCOUNTERS AT A CONFERENCE: The information used here was provided by Richard McGehee (as in 3.6) and by John Mather in 1993 at the Oberwolfach meeting mentioned in note 1.14. The paper of Mather and McGehee is [M & M,1975]. Sheldon's paper is [Sh,1978].

3.9. FROM FOUR TO FIVE BODIES: This section is based on a discussion between Gerver and FD at the Midwest Dynamical Systems Conference at Northwestern University in March 1991. The Riemann conjecture proved by Gerver in 1969, as a graduate student, states that the function $\sum_{n=1}^{\infty}$ $(\sin n^2x)/n^2)$ is nowhere differentiable. FD learned about the proof of this conjecture (which was unsuccessfully attempted by Weierstrass) not from Gerver but from Norbert Schlomiuk of the University of Montreal during the Canadian Mathematical Society Winter Meeting, in December 1991, at the University of Victoria. Gerver's heuristic example of a solution leading to a noncollision singularity was published in [Ge,1984].

3.10. THE END OF A CENTURY'S QUEST: This section is based on discussions FD had with Jeff Xia at Northwestern University (as in 3.9) and at the 1995 Seattle meeting noted in 3.4; with John Mather (as in 3.8); and the exchange of E-mail messages with Don Saari (as in 3.4). A short biography of Xia appeared in the Notices of the American Mathematical Society **40**, 1993, November issue, pp. 1220–1221, together with the announcement of the Blumenthal Prize. Xia's paper is [X,1992]. Also see the expository article [S & X,1995] referred to in 3.0.

3.11. A SYMMETRIC DIGRESSION: Some of the classical results on the isosceles problem can be found in [Win,1941]. Other relevant papers and books are: Wilczynski [Wi,1913], Sitnikov [Sit,1961], Moser [Mos,1973], Devaney [De,1980, 1982], Moeckel [Mo,1984], Broucke [Bro,1979], Simó [Sim,1981], and Simó and Martinez [S & M,1988]. This list is far from complete.

3.12. AN IDEA AT DINNER: Gerver's example of a planar solution leading to a noncollision singularity is given in [Ge,1991], where the discussion between Gerver and Scott Brown is also described. The story of Williams's insight into the problem was told to FD by Richard McGehee in Berkeley, as in 3.6.

Chapter 4.
Stability

4.0. The epigraph to chapter 4 is taken from [Mos,1978].

4.1. A LONGING FOR ORDER: The introductory story of Laplace's fruitless call at the home of d'Alembert and the subsequent communication of his essay on the general principles of mechanics are described in [Be,1937] and [Nm,1960]. We have imagined the scene in d'Alembert's house. [Sz,1984] reviews the many different notions of stability used in celestial mechanics. Aristotle's and Archimedes' contributions are described in [M,1959].

4.2. THE MARQUIS AND THE EMPEROR: See [Be,1937]. A brief historical account of the development of perturbation and averaging methods can be found in [V,1984]. Many of the fundamental ideas appear in Laplace's *Mécanique céleste* [Lap,1799–1825], but the first clear exposition appears to have been that of Lagrange, in *Mécanique analytique* [Lag,1788], as described in section 4.3. The story of Adams and Le Verrier is told in [Pet,1993]. A translation of Laplace's essay "Concerning Probability" can be found in [Nm,1960], vol. 2, along with a biographical essay on Laplace, from which some details of Laplace's life are taken. Lagrange's remark on God is quoted from this source.

4.3. MUSIC OF THE SPHERES: For this section we appeal to [Co,1990], [Cal,1982], [Nm,1960], and [Be,1937]. Also see [V,1984], [Koe,1986], and the note above. Lagrange's words on his deathbed were allegedly spoken to his friends Monge, Lacépède, and Chaptal. They were reconstructed, from notes taken by the latter, in M. Delambre's biographical essay, pages xliv–xlv of [Lag,1867–92], vol. 1; cf. [Koe,1986].

4.4. ETERNAL RETURN: The sources for this section are [Co,1990], [Be,1937], and [Po,1993]. The recurrence result appears as Theorem I in section 8 of [Po, 1890].

4.5. PERTURBING THE WORLD: There are several books in Romanian about Spiru Haretu, e.g., [Bâ,1972], [Din,1970], and [Or,1976]. Two recent articles analyzing his contribution to celestial mechanics are [R,1985] and [Pá,1991]. Haretu's thesis is eulogistically cited by Poincaré in [Po,1905]. In fact, Haretu was not only the first Romanian, but the first foreigner granted a Ph.D. in mathematics in Paris, as noted in a letter from the dean of the Faculty of Science to the Romanian minister of education, Gheorge Chiţe. The book on "social mechanics" is [Har,1910]. Meffroy's study is [Me,1958].

4.6. HOW STABLE IS STABLE?: Poincaré's statements can be found in [Po,1898]. An English translation appears in Goroff's introduction to [Po,1993].

4.7. THE QUALITATIVE AGE: Bruns's crucial paper is [Bru,1887]. In the Western literature, Liapunov's name appears in at least five different spellings: Liapunov, Lyapunov, Liapunoff, Lyapunoff, and Ljapunow. We adopt the first. Liapunov's papers mentioned here can be found in French and English translations as [Li,1947], [Li,1966], [Li,1992]. A biography of Liapunov and a bibliography of his works have appeared in [Smi,1992] and [B,1992]. A note describing the events surrounding Liapunov's death was published by Wilson [Wi,1994].

4.8. LINEARIZATION AND ITS LIMITS: The papers by Grobman and Hartman are [Gr,1959, 1962] and [Ha,1960, 1963] (in fact, the analytic case of the Hartman-Grobman theorem had already been proved by Poincaré in 1879; see [Po,1951–6]). Most classical textbooks on the qualitative theory of differential equations contain the main results of the theory of stability developed by Liapunov. See [H & S,1974], for example. Liapunov exponents for arbitrary solutions were first justified in [O,1968]; an introduction can be found in [G & H,1983].

4.9. THE STABILITY OF MODELS: Introductory descriptions of structural stability can be found in [H & S,1974] and [G & H,1983].

4.10. PLANETS IN BALANCE: The papers of L. Euler and J. L. Lagrange are [Eu,1767] and [L,1873]. Details on the problem of the number of central configurations can be found in [Win,1941]. Numerous people have worked on this problem, including Dziobek and Moulton at the turn of the century, and more recently Alain Albouy, Greg Buck, Josefina Casasayas, Alain Chenciner, Nelly Fayçal, G. R. Hall, Jaume Llibre, Chris McCord, Ken Meyer, Rick Moeckel, Ana Nunes, Filomena Pacella, Julian Palmore, Don Saari, Dieter Schmidt, Carles Simó, Stephen Smale, Nick Tien, and Jeff Xia. Among their many papers are: [Al,1995], [Bu,1990, 1991], [CLN,1994], [Dz,1900], [Ll,1991], [Mo,1985, 1990], [Mou,1910], [P,1986, 1987], [Pal,1974, 1975a, 1975b, 1982], [Sim,1977], [X,1991]. Sundman's paper is [Su,1912]. Saari's paper is [S,1980]. For a more complete source of references on this subject, see [Ll,1991]. A new approach to central configurations was initiated in the papers of Smale [Sm,1970a, 1970b].

Chapter 5.
KAM Theory

5.0. The epigraph to chapter 5 is taken from [Bi,1920].

5.1. SIMPLIFY AND SOLVE: Kolmogorov's 1954 ICM lecture in Amsterdam originally appeared in Russian (under a French title) as [K,1957]. The English translation of the lecture is [K,1991a]. Short biographies of Kolmogrov have been published by V. M. Tikhomirov and P. S. Alexandrov in [T,1991–92]. An obituary of Kolmogorov was published by Arnold [Ar,1989b]. An interesting article by Arnold entitled "On A. N. Kolmogorov," in preparation for the American Mathematical Society, was shown to FD by Smilka Zdravkovska in 1993; it appeared in [Z & D,1994]. Some of our information derives from this, and some from a handwritten account that V. I. Arnold provided at FD's request. Borges's story "Funes the Memorius" appears in [Bo,1964], in which a second story, "The Library of Babel," explores the notions of infinity and randomness.

5.2. QUASI-PERIODIC MOTIONS: There are several sources concerning KAM theory and the small denominator problem, but no book (yet!) is exclusively dedicated to it. For the interested (and mathematically adept) reader, we suggest [Ar,1983, 1989a] and [Ga,1983].

5.3. PERTURBING THE TORI: See 5.2 above.

5.4. LETTERS, A LOST SOLUTION, AND POLITICS: For this section we used [Be,1937], [Bö,1992], the brief biographies of Weierstrass in [Cal,1982] and of Kovalevskaia in [EB,1986], Goroff's introduction to [Po,1993], and the sources mentioned in 5.1. For extracts of correspondence on King Oscar's Prize, see [B-G,1994] (cf. 1.7). The paper of Krylov and Bogoliubov is [K & B,1937]. Yasha Sinai provided some of the information on Kolmogorov and his seminar in discussions with PH at Princeton in January 1995. Lysenko (1998–1976) rose to prominence in Soviet science through his spurious genetic theories, which promised enormous crop yields through interbreeding. He was scientific director and later director of the All Union Institute of Selection and Genetics in Odessa in the 1930s and subsequently held influential posts in the Academy of Sciences and the All Union Academy of Agricultural Sciences. The controversial papers on Mendelian genetics are [K,1940], [Ly,1940], and [Ko,1940].

5.5. WORRYING AT THE PROOF: This section is based on several conversations between FD and Jürgen Moser in Oberwolfach in July 1993 (as in 1.14), on

[Ze,1993], and on E-mail correspondence with Moser in May 1995. The review of Kolmogorov's paper by Moser is [Mos,1959]. The book of Siegel and Moser appeared in English as [S & M,1971]. Moser's paper containing the twist theorem is [Mos,1962]. Some details on Moser's meeting with Kolmogorov and Arnold and his visit to Moscow derive from the conversation with Sinai noted in 5.4.

5.6. TWIST MAPS: See [Mos,1962,1973] and [Ar,1989a].

5.7. A GIFTED STUDENT: This section is based on [Z,1987], from which Arnold's remarks are quoted, and the literature on KAM theory mentioned in 5.2. In Montreal, on several occasions between 1985 and 1993, Leon Glass told PH about Arnold's work in mathematical biology (cf. [G & M,1988]). A few details derive from Arnold's handwritten account, as mentioned in 5.1, and PH's discussion with Sinai noted in 5.4. Arnold's announcement and proof of the "analytic" KAM theorem appeared in [Ar,1962, 1963a,1963b].

5.8. CHAOS DIFFUSES: The sources of 5.7 were again used in this section. The first paper on diffusion is [Ar,1964]. [H & M,1983a, 1983b] are early references in the Western literature on diffusion in slightly more general classes of Hamiltonian systems. Xia's papers on the existence of Arnold diffusion in the three-body problem are [X,1992, 1993]; Chierchia and Gallavotti's work on oblate planetary rotation is [C & G,1994]. An introductory account on this subject with a discussion on the astronomical implications of Arnold diffusion, chaos in the solar system, and the Kirkwood gaps appears in [B & D,1993].

5.9. EPILOGUE: The books of Glass & Mackey, Gleick, Lorenz, Ruelle, and Stewart are [G & M,1988], [Gl,1987], [Lo,1993], [Ru,1991], and [St,1989]. The volume [HM & S,1993] contains the proceedings of the conference that celebrated Smale's sixtieth birthday. It includes a survey of Smale's work, and further details appear in [Sm,1991].

Bibliography

[Al,1995] Albouy, A. Symétrie des configurations centrales de quatre corps. *Comptes Rendus Acad. Sci. Paris, Série I*, **320** (1995), 217–220.

[A,1968a] Alekseev, V. M. Quasirandom dynamical systems. I. Quasirandom diffeomorphisms. *Mathematics USSR Sbornik* **5**, 1 (1968), 73–128.

[A,1968b] Alekseev, V. M. Quasirandom dynamical systems. II. One-dimensional nonlinear oscillations in a field with periodic perturbation. *Mathematics USSR Sbornik* **6**, 4 (1968), 505–560.

[A,1969] Alekseev, V. M. Quasirandom dynamical systems. III. Quasirandom oscillations of one-dimensional oscillators. *Mathematics USSR Sbornik* **7**, 1 (1969), 1–43.

[And,1994] Andersson, K. G. Poincaré's discovery of homoclinic points. *Archive for History of Exact Sciences* **48** (1994), 133–147.

[A & P,1937] Andronov, A. A., and Pontryagin, L. Systèmes grossieres. *Dokl.-Akad.Nauk.SSSR* **14** (1937), 247–251.

[Ar,1962] Arnold, V. I. The classical theory of perturbations and the problem of stability of planetary systems. *Soviet Mathematics Doklady* **3** (1962), 1008–1012.

[Ar,1963a] Arnold, V. I. Proof of A. N. Kolmogorov's theorem on the preservation of quasiperiodic motions under small perturbations of the Hamiltonian. *Russian Mathematical Surveys* **18** (5) (1963), 9–36.

[Ar,1963b] Arnold, V. I. Small divisor problems in classical and celestial mechanics. *Russian Mathematical Surveys* **18** (6) (1963), 85–192.

[Ar,1964] Arnold, V. I. Instability of dynamical systems with several degrees of freedom. *Soviet Mathematics Doklady* **5** (1964), 342–355.

[Ar,1973] Arnold, V. I. *Ordinary Differential Equations*. The MIT Press, Cambridge, Mass., 1973.

[Ar,1983] Arnold, V. I. *Geometrical Methods in the Theory of Ordinary Differential Equations*. Springer-Verlag, New York–Heidelberg–Berlin, 1983.

[Ar,1988] Arnold, V. I., ed. *Dynamical Systems III*. Springer-Verlag, New York–Heidelberg–Berlin, 1988.

[Ar,1989a] Arnold, V. I. *Mathematical Methods of Classical Mechanics*. Springer Verlag, New York–Heidelberg–Berlin, 1989.

[Ar,1989b] Arnold, V. I. A. N. Kolmogorov. *Physics Today*, October 1989, 148–149.

[B & D,1993] Bakker, L. F., and Diacu, F. N. On the existence of celestial bodies with unpredictable motion in the solar system and on the Kirkwood gaps. *Romanian Astronomical Journal* **3**, 2 (1993), 139–155.

[Bâ,1972] Bâldescu, E., *Spiru Haret în ştiinţă, filozofie, politică, pedagogie, învăţămînt*, Editura didactică şi pedagogică, Bucureşti, 1972.

[Ba & D,1993] Ballinger, G., and Diacu, F. N. Collision and near collision orbits in post-Newtonian gravitational systems. *Romanian Astronomical Journal* **3**, 1 (1993), 51–59.

[Bet al.,1992] Banks, J.; Brooks, J.; Cairns, G.; Davis, G.; and Stacey, P. On Devaney's definition of chaos. *American Mathematical Monthly*, April 1992, 332.

[B,1992] Barret, J. F. Bibliography of A. M. Lyapunov's work. *International Journal of Control* **55**, 3 (1992), 785–790.

[B-G,1994] Barrow-Green, J. Oscar II's prize competition and the error in Poincaré's memoir on the three body problem. *Archive for History of Exact Sciences* **48** (1994), 107–131.

[Be,1937] Bell, E. T. *Men of Mathematics*. Simon and Schuster, New York, 1937.

[Ber,1978] Berry, M. V. Regular and irregular motion. In *Topics in Nonlinear Dynamics*, pp. 16–120. AIP Conference Proceedings **46**, ed. Siebe Jorna. American Institute of Physics, New York, 1978.

[Bi,1913] Birkhoff, G. D. Proof of Poincaré's geometric theorem. *Transactions of the American Mathematical Society* **14** (1913), 14–22.

[Bi,1920] Birkhoff, G. D. Recent advances in dynamics. *Science (New Series)* **51** (1920), 51–55.

[Bi,1927] Birkhoff, G. D. *Dynamical Systems*. American Mathematical Society, Providence, R.I., 1927. (Reprinted with an introduction by J. Moser and a preface by M. Morse, 1966.)

[Bi,1935] Birkhoff, G. D. Nouvelles recherches sur les systèmes dynamiques. *Mémoriae Pont. Acad. Sci. Novi Lyncaei* **1** (1935), 85–216.

[Bi,1968] Birkhoff, G. D. *Collected Mathematical Papers*. American Mathematical Society, Providence, R.I., 1950. (Reprinted by Dover, New York, 1968.)

[Bö,1992] Bölling R. . . . Deine Sonia: A reading from a burned letter. *The Mathematical Intelligencer* **14**, 3 (1992), 24–30.

[Bo,1964] Borges, J. L. *Labyrinths: Selected Stories and Other Writings*. New Directions, New York, 1964.

[Br,1994] Bressoud, D. M. Review of "The Search for E. T. Bell, also known as John Taine," by Constance Reid. *The Mathematical Intelligencer* **16**, 3 (1994), 72–74.

[Bro,1979] Broucke, R. On the isosceles triangle configuration in the planar general three-body problem. *Astronomy and Astrophysics* **73** (1979), 303–313.

[Bru,1887] Bruns, H. Über die Integrale des Vielkörper-Problems. *Acta Mathematica* **11** (1887), 25–96.

[Bch,1904] Buchholz, H. Poincarés Preisarbeit von 1889/90 und Gyldéns Forschung über das Problem der drei Körper in ihren Ergebnissen für die Astronomie. *Physikalische Zeitschrift* **7**, 5 (1904), 180–186.

[Bu,1990] Buck, G. On clustering in central configurations. *Proceedings of the American Mathematical Society* **108**, 3 (1990), 801–810.

[Bu,1991] Buck, G. The collinear central configuration of *n* equal masses. *Celestial Mechanics* **51** (1991), 305–317.

[Cab,1988] Cabral, H. The masses in an isosceles solution of the three-body problem. *Celestial Mechanics* **41** (1988), 175–177.

[Cal,1982] Calinger, R., ed. *Classics of Mathematics*. Moore Publishing Company, Oak Park, Ill., 1982.

[C & L,1945] Cartwright, M. L., and Littlewood, J. E. On nonlinear differential equations of the second order, I: The equation: $\ddot{y} + k(1 - y^2) + \dot{y} + y = b\,\lambda\,k\cos(\lambda\,t + a)$, k large. *J. London Math. Soc.* **20** (1945), 180–189.

[Ca,1972] Cartwright, M. L. The early history of the theory of dynamical systems and its relationship to the theory of nonlinear oscillations. In *Proceedings of the Colloquium on Smooth Dynamical Systems*, I.1–I.13. Department of Mathematics, University of Southampton, U.K., 18–22 September 1972.

[CLN,1994] Casasayas, J.; Llibre, J.; and Nunes, A. Central configurations of the planar $1 + N$ body problem. *Celestial Mechanics* **60** (1994), 273–288.

[Ch,1920] Chazy, J. Sur les singularités impossible du problème des *n* corps. *Comptes Rendus Hebdomadaires des Séances de l'Academie des Sciences de Paris* **170** (1920), 575–577.

[C & G,1994] Chierchia, L., and Gallavotti, G. Drift and diffusion in phase space. *Annales de l' Institut Poincaré* B, **60** (1994), 1–144.

[Co,1990] *Collier's Encyclopedia*. Macmillan, New York, 1990.

[D,1914] Darboux, G. Éloge historique d'Henri Poincaré. *Mem. Acad. Sci. Inst. Fr.* **52**, 74 (1914).

[Da,1900] Darwin, G. President's address. *Monthly Notices of the Royal Astronomical Society* **60** (1900), 412.

[D & H,1981] Davis, P. J., and Hersh, R. *The Mathematical Experience*. Birkhäuser, Boston, 1981.

[De,1980] Devaney, R. Triple collision in the planar isosceles three-body problem. *Inventiones Mathematicae* **60** (1980), 249–267.

[De,1982] Devaney, R. Motion near total collapse in the planar isosceles three-body problem. *Celestial Mechanics* **28** (1982), 25–36.

[De,1986] Devaney, R. *An Introduction to Chaotic Dynamical Systems*. Benjamin Cummings, Menlo Park, Calif., 1986.

[Di,1992a] Diacu, F. N. Regularization of partial collisions in the *N*-body problem. *Differential and Integral Equations* **5** (1992), 103–136.

[Di,1992b] Diacu, F. N. *Singularities of the N-Body Problem*. Les Publications CRM, Montreal, 1992.

[Di,1993a] Diacu, F. N. The planar isosceles problem for Maneff's gravitational law. *Journal of Mathematical Physics* **34**, 12 (1993), 5671–5690.

[Di,1993b] Diacu, F. N. Painlevé's conjecture. *The Mathematical Intelligencer* **15**, 2 (1993), 6–12.

[Di,1996a] Diacu, F. N. On the Mücket-Treder gravitational law. In *Proceedings of the International Conference on Hamiltonian Systems and Celestial Mechanics*, ed. E. A. Lacomba and J. Llibre. Cocoyoc, Mexico, 1994 (to appear, 1996).

[Di,1996b] Diacu, F. N. Near-collision dynamics for particle systems with quasi-homogeneous potential laws. *Journal of Differential Equations* (to appear, 1996).

[Din,1970] Dinu, C., *Spiru Haret*, Editura didactică şi pedagogică, Bucureşti, 1970.

[Dy,1995] Dyson, F. The scientist as rebel. *New York Review of Books* 42, 9 (May 25, 1995), 31–33.

[Dz,1900] Dziobek, O. Ueber einen merkwürdigen Fall des Vielkörperproblems. *Astronomische Nachrichten* **152** (1900), 34–46.

[E,1971] Easton, R. Regularization of vector fields by surgery. *Journal of Differential Equations* **10** (1971), 92–99.

[Ek,1988] Ekeland, I. *Mathematics and the Unexpected.* University of Chicago Press, Chicago, 1988.

[El,1990] Elbialy, M. S. Collision singularities in celestial mechanics. *SIAM Journal of Mathematical Analysis* **21** (1990), 1563–1593.

[EB,1986] *Encyclopaedia Britannica.* 15th ed. Encyclopaedia Britannica, Inc., Chicago, 1986.

[Eu,1767] Euler, L. De moto rectilineo trium corporum se mutuo attrahentium. *Novo Comm. Acad. Sci. Imp. Petrop.* **11** (1767), 144–151.

[F,1972] Fang, J. *Mathematicians from Antiquity to Today.* Paideia Press, Hauppauge, N.Y., 1972.

[Ga,1983] Gallavotti, G. *The Elements of Mechanics.* Springer-Verlag, New York–Heidelberg–Berlin, 1983.

[Ge,1984] Gerver, J. A possible model for a singularity without collisions in the five-body problem. *Journal of Differential Equations* **52** (1984), 76–90.

[Ge,1991] Gerver, J. The existence of pseudocollisions in the plane. *Journal of Differential Equations* **89** (1991), 1–68.

[G & M,1988] Glass, L., and Mackey, M. C. *From Clocks to Chaos.* Princeton University Press, Princeton, N.J., 1988.

[Gl,1987] Gleick, J. *Chaos—Making a New Science.* Penguin Books, New York, 1987.

[Gr,1959] Grobman, D. M. Homeomorphisms of systems of differential equations. *Dokl. Akad. Nauk. SSSR* **128** (1959), 880–881.

[Gr,1962] Grobman, D. M. Topological classification of the neighborhood of a singular point in n-dimensional phase space. *Mat. Sbornik (New Series)* **56** (98) (1962), 77–94.

[G & H,1983] Guckenheimer, J., and Holmes, P. *Nonlinear Oscillations, Dynamical Systems and Bifurcations of Vector Fields.* Springer-Verlag, New York–Heidelberg–Berlin, 1983. (Fourth corrected printing, 1994.)

[Gy,1887] Gyldén, H. Untersuchungen über die Convergenz der Reihen, welche zur Darstellung der Coordinaten der Planeten angewendet werden. *Acta Mathematica* **9** (1887), 185–294.

[H,1975] Hagihara, Y. *Celestial Mechanics.* Vol. 2, part 1. The MIT Press, Cambridge, Mass., 1975.

[Har,1910] Haretu, S. *Mécanique sociale.* Gauthier-Villars, Paris, 1910.

[Ha,1960] Hartman, P. A lemma in the theory of structural stability of differential equations. *Proceedings of the American Mathematical Society* **11** (1960), 610–620.

[Ha,1963] Hartman, P. On the linearization of differential equations. *Proceedings of the American Mathematical Society* **14** (1963), 568–573.

[Hay,1990] Hayles, K. *Chaos Bound: Orderly Disorder in Contemporary Literature and Science.* Cornell University Press, Ithaca, N.Y., 1990.

[Hi,1984] Hirsch, M. W. The dynamical systems approach to differential equations. *Bulletin of the American Mathematical Society* **11** (1984), 451–514.

[HM & S,1993] Hirsch, M. W.; Marsden, J. E.; and Shub, M., eds. *From Topology to Computation: Proceedings of the Smalefest*. Springer-Verlag, New York–Heidelberg–Berlin, 1993.

[H & S,1974] Hirsch, M. W., and Smale, S. *Differential Equations, Dynamical Systems and Linear Algebra*. Academic Press, New York, 1974.

[Ho,1990] Holmes, P. Poincaré, celestial mechanics, dynamical systems theory and chaos. *Physics Reports* **193**, 3 (1990), 137–163.

[H & M,1983a] Holmes, P., and Marsden, J. E. Melnikov's method and Arnold diffusion for perturbations of integrable Hamiltonian systems. *J. Math. Phys.* **23** (1983), 669–675.

[H & M,1983b] Holmes, P., and Marsden, J. E. Horseshoes and Arnold diffusion for Hamiltonian systems on Lie groups. *Indiana U. Math. J.* **32** (1983), 273–310.

[Hop,1942] Hopf, E. Abzweigung einer periodischen Lösung von einer stationären Lösung eines Differential-Systems. *Berichte Math.-Phys. Kl. Sächs. Akad. Wiss. Leipzig* **94** (1942), 1–22, and *Berichte Math.-Nat. Kl. Vreh. Sächs. Akad. Wiss. Leipzig* **95**, 1 (1942), 3–22.

[H & Y,1993] Hunt, B. R., and Yorke, J. A. Maxwell on Chaos. *Nonlinear Science Today* **3**, 1 (1993), 1, 3–4.

[Koe,1986] Koetsier, T. Joseph Louis Lagrange (1786–1813), his life, his work and his personality. *Nieuw Archief voor Wiskunde* **3**, 4 (1986), 191–206.

[Ko,1940] Kolman, E. Is it possible to prove or disprove Mendelism by mathematical and statistical methods? *Dokl. Akad. Nauk. SSSR* **28**, 9 (1940), 834–838 (in Russian: English translation published in *Comptes Rendus [Doklady] de l'Academie des Sciences de l'URSS*).

[K,1940] Kolmogorov, A. N. On a new confirmation of Mendel's laws. *Dokl. Akad. Nauk. SSSR* **27**, 1 (1940), 38–42 (in Russian); English translation in *Comptes Rendus [Doklady] de l'Academie des Sciences de l'URSS*). Reprinted in *Selected Works of A. N. Kolmogorov*, vol. 2, pp. 222–227, ed. V. M. Tikhomirov. Kluwer, Dordrecht–Boston–London, 1991.

[K,1957] Kolmogorov, A. N. Théorie générale des systèmes dynamiques et mécanique classique. In *Proceedings of the International Congress of Mathematicians*, Amsterdam, 1954, vol. 1, pp. 315–333. Erveb P. Noordhoff N. V., Groningen; North-Holland Publishing Co., Amsterdam, 1957.

[K,1991a] Kolmogorov, A. N. The general theory of dynamical systems and classical mechanics. In *Selected Works of A. N. Kolmogorov*, vol. 1, pp. 355–374, ed. V. M. Tikhomirov. Kluwer, Dordrecht–Boston–London, 1991.

[K,1991b] Kolmogorov, A. N. On the preservation of conditionally periodic motions under small variations of the Hamilton function. In *Selected Works of A. N. Kolmogorov*, vol. 1, pp. 349–354, ed. V. M. Tikhomirov. Kluwer, Dordrecht–Boston–London, 1991.

[K & B,1937] Kryloff, N., and Bogolyuboff, N. La théorie générale de la mesure dans son application à l'étude des systèmes dynamiques de la mécanique non linéaire. *Annals of Mathematics* **38**, 1 (1937), 65–113.

[L,1873] Lagrange, J. L. *Oeuvres*. Vol. 6, pp. 272–292. Paris, 1873.

[Lag,1788] Lagrange, J. L. *Mécanique analytique*. 2 vols. Desaint, Paris, 1788. (Reprinted by Librairie Albert Blanchard, Paris, 1873, and as vols. 11 and 12 of [Lag,1867–92].)

[Lag,1867–92] Lagrange, J. L. *Oeuvres*. 14 vols. Ed. J. A. Serret and G. Darboux. Paris, 1867–1892.

[Lap,1799–1825] Laplace, P. S. *Traité de mécanique céleste*. 5 vols. Paris, 1799–1825. (Reprinted as vols. 1–5 of [Lap,1878].)

[Lap,1878–1912] Laplace, P. S. *Oeuvres*. 14 vols. Gauthier-Villars, Paris, 1878–1912.

[Levi,1981] Levi, M. Qualitative analysis of the periodically forced relaxation oscillations. *Memoirs of the American Mathematical Society* **214** (1981), 1–147.

[Le,1920] Levi-Civita, T. Sur la régularisation du problème des trois corps. *Acta Mathematica* **42** (1920), 99–144.

[Levin,1949] Levinson, N. A second-order differential equation with singular solutions. *Annals of Mathematics* **50** (1949), 127–153.

[Lev,1968] Levy, J. La contribution française au développement de la mécanique céleste au cours de trois derniers siècles. In *Colloque sur le problème des n corps*, pp. 13–20. Editions du Centre Nationale de la Rechrche Scientifique, Paris VII, 1968.

[Li,1947] Liapunov, M. A. Problème général de la stabilité de mouvement. *Annals of Mathematical Studies* **17**. Princeton University Press, Princeton, N.J., 1947.

[Li,1966] Liapunov, M. A. *Stability of Motion*. Academic Press, New York and London, 1966.

[Li,1992] Liapunov, M. A. The general problem of the stability of motion. *International Journal of Control* **55**, 3 (1992), 531–773.

[Lit,1986] Littlewood, J. E. *Littlewood's Miscellany*. Edited by Béla Bollobás. Cambridge University Press, Cambridge, U.K., 1986.

[Ll,1991] Llibre, J. On the number of central configurations in the *n*-body problem. *Celestial Mechanics* **50** (1991), 89–96.

[Ló,1993] Lorenz, E. N. *The Essence of Chaos*. University of Washington Press, Seattle, 1993.

[Ly,1940] Lysenko, T. D. In response to an article by A. N. Kolmogorov. *Dokl. Akad. Nauk. SSSR* **28**, 9 (1940), 832–833 (in Russian: English translation published in *Comptes Rendus [Doklady] de l'Academie des Sciences de l'URSS*).

[M,1959] Magnus, K. Zur Entwicklung des Stabilitätsbegriffes der Mechanik. *Naturwissenschaften* **46** (1959), 590–595.

[Ma,1924] Maneff, G. La gravitation et le principe de l'égalité de l'action et de la réaction. *Comptes Rendus Acad. Sci. Paris* **178** (1924), 2159–2161.

[Ma,1925] Maneff, G. Die Gravitation und das Prinzip von Wirkung und Gegenwirkung. *Zeitschrift für Physik* **31** (1925), 786–802.

[Ma,1930a] Maneff, G. Le principe de la moindre action et la gravitation. *Comptes Rendus Acad. Sci. Paris* **190** (1930), 963–965.

[Ma,1930b] Maneff, G. La gravitation et l'énergie au zéro. *Comptes Rendus Acad. Sci. Paris* **190** (1930), 1374–1377.

[M & M,1975] Mather, J., and McGehee, R. Solutions of the collinear four-body problem which become unbounded in finite time. In *Dynamical Systems Theory and Applications*, pp. 573–587, ed. J. Moser. Lecture Notes in Physics. Springer-Verlag, New York–Heidelberg–Berlin, 1975.

[Mc,1974] McGehee, R. Triple collision in the collinear three-body problem. *Inventiones Mathematicae* **27** (1974), 191–227.

[Mc,1975] McGehee, R. Triple collision in Newtonian gravitational systems. In *Dynamical Systems Theory and Applications*, pp. 550–572, ed. J. Moser. Lecture Notes in Physics. Springer-Verlag, New York–Heidelberg–Berlin, 1975.

[Mc,1981] McGehee, R. Double collisions for a classical particle system with nongravitational interactions. *Commentarii Mathematici Helvetici* **56** (1981), 524–557.

[Mc,1986] McGehee, R. Von Zeipel's theorem on singularities in celestial mechanics. *Expositionae Mathematicae* **4** (1986), 335–345.

[Me,1958] Meffroy, J. Sur l'existence effective du terme séculaire pur de la perturbation du troisième ordre des grands axes. *Bull. Astron.* **21** (1958), 261–322.

[Mey,1994] Meyer, M. *The richest vagabond* (a review of "Alfred Nobel: A Biography," by Kenne Fant. *New York Review of Books* **41** (1 and 2) (January 13, 1994), 26–27.

[Mi,1972] Mihăileanu, N. *Istoria Matematicii*. Vol. 1. Editura Ştiinţifică şi Enciclopedică, Bucharest, 1972.

[Mi,1981] Mihăileanu, N. *Istoria Matematicii*. Vol. 2. Editura Ştiinţifică şi Enciclopedică, Bucharest, 1981.

[Mi & Mr,1995] Mischiakow, K., and Mrozek, M. Chaos in the Lorenz equations: A computer-assisted proof. *Bulletin of the American Mathematical Society* **32**, 1 (1995), 66–72.

[Mo,1984] Moeckel, R. Heteroclinic phenomena in the isosceles three-body problem. *SIAM Journal of Mathematical Analysis* **15**, 5 (1984), 857–876.

[Mo,1985] Moeckel, R. Relative equilibria of the four-body problem. *Ergodic Theory and Dynamical Systems* **5** (1985), 417–435.

[Mo,1990] Moeckel, R. On central configurations. *Mathematische Zeitschrift* **205** (1990), 499–517.

[Mor,1938] Morse, M. *Symbolic Dynamics*. Notes by R. Oldenburger of Lectures by Marston Morse. The Institute for Advanced Study, Princeton, N.J., 1938.

[Mor,1946] Morse, M. George David Birkhoff and his mathematical work. *Bulletin of the American Mathematical Society* **52** (1946), 357–391.

[Mos,1959] Moser, J. K. Review of Kolmogorov, A. N., "Théorie générale des systèmes dynamiques et mécanique classique." *Mathematical Reviews* **20**, 6 (1959), 675–676.

[Mos,1962] Moser, J. K. On invariant curves of area-preserving mappings of an annulus. *Nachr. Akad. Wiss. Göttingen II, Math. Phys. Kl.* 1962, 1–20.

[Mos,1973] Moser, J. K. *Stable and Random Motions in Dynamical Systems*. Princeton University Press, Princeton, N.J., 1973.

[Mos,1978] Moser, J. K. Is the solar system stable? *The Mathematical Intelligencer* **1**, 2 (1978), 65–71.

[Mou,1910] Moulton, F. R. The straight line solutions of the problem of *n* bodies. *Annals of Mathematics* **12** (1910), 1–17.

[Mou,1912] Moulton, F.R. M. Henri Poincaré. *Popular Astronomy* **10**, 10 (1912), 625.

[Mou,1914] Moulton, F. R. *An Introduction to Celestial Mechanics*. Macmillan, London, 1914. (Reprinted by Dover, New York, 1970.)

[Nm,1960] Newman, J. R., ed. *The World of Mathematics. Vols. 1–4.* George Allen and Unwin, London, 1960.

[N,1686] Newton, I. *Philosophiae Naturalis Principia Mathematica.* S. Pepys, Royal Society Press, London, 1686.

[N,1934] Newton, I. *Sir Isaac Newton's Mathematical Principles of Natural Philosophy and His System of the World.* 2 vols. Trans. Andrew Motte (1729). Revised, with historical and explanatory appendix, by Florian Cajori. University of California Press, Berkeley, 1934.

[Or,1976] Orăscu, Ş., *Spiru Haret*, Editura ştiinţifică şi enciclopedică, Bucureşti, 1976.

[O,1968] Oseledec, V. I. A multiplicative ergodic theorem: Liapunov characteristic numbers for dynamical systems. *Trans. Moscow Math. Soc.* **19** (1968), 197–231.

[P,1986] Pacella, F. Equivariant Morse theory for flows and an application to the n-body problem. *Transactions of the American Mathematical Society* **297** (1986), 41–52.

[P,1987] Pacella, F. Central configurations of the n-body problem via equivariant Morse theory. *Archive for Rational Mechanics and Analysis* **97** (1987), 59–73.

[Pa,1897] Painlevé, P. *Leçons sur la théorie analytique des équations differentielles.* Hermann, Paris, 1897.

[Pa,1972] Painlevé, P. *Oeuvres.* Vol 1. Ed. Centr. Nat. Rech. Sci., Paris, 1972.

[Pá,1991] Pál, Á. Spiru Haretu's theorem. *Romanian Astronomical Journal* **1**, 1–2 (1991), 5–11.

[Pal,1974] Palmore, J. I. Classifying relative equilibria. I. *Bulletin of the American Mathematical Society* **79**, 5 (1973), 904–908.

[Pal,1975a] Palmore, J. I. Classifying relative equilibria. II. *Bulletin of the American Mathematical Society* **81**, 2 (1975), 589–591.

[Pal,1975b] Palmore, J. I. Classifying relative equilibria. III. *Letters in Mathematical Physics* **1** (1975), 71–73.

[Pal,1982] Palmore, J. I. Collinear relative equilibria of the planar n-body problem. *Celestial Mechanics* **28** (1982), 17–24.

[Pan,1951] Pancoe, A. "Properties of the solution of the second Painlevé transcendent." M.Sc. thesis, Northwestern University, 1982.

[Pe,1995] Peetre, J. Outline of a scientific biography of E. Meissel (1826–1895). *Historia Mathematica 22* (1995), 154–78.

[Pet,1993] Peterson, I. *Newton's Clock—Chaos in the Solar System.* Freeman, New York, 1993.

[Po,1881–6] Poincaré, H. J. Mémoire sur les courbes définies par une équation différentielle. *Journal de Mathématiques 7* (1881), 375–422, and **8** (1882), 251–296; *Journal de Mathématiques Pures et Appliquées* **1** (1885), 167–244, and **2** (1886), 151–217.

[Po,1890] Poincaré, H. J. Sur le problème des trois corps et les équations de la dynamique. *Acta Mathematica* **13** (1890), 1–270.

[Po,1891] Poincaré, H. J. Sur le problème des trois corps. *Bulletin Astronomique* **8** (1891), 12–24.

[Po,1892–3-9] Poincaré, H. J. *Les méthodes nouvelles de la de mécanique céleste.* *Vols. 1–3.* Gauthier-Villars, Paris, 1892, 1893, 1899. (Reprinted by Librairie Albert Blanchard, Paris, 1987.)

[Po,1898] Poincaré, H. J. Sur la stabilité du système solaire. In *Annuaire pour l'an 1898 par le Bureau des Longitudes*, pp. B. 1–2. Gauthier-Villars, Paris, 1898.

[Po,1905] Poincaré, H. J. *Leçons de mécanique céleste.* Gauthier-Villars, Paris, 1905.

[Po,1912] Poincaré, H. J. Sur un théorème de géométrie. *Rendiconti del Circolo Matematico di Palermo* **33** (1912), 375–407.

[Po,1929] Poincaré, H. J. *The Foundation of Science.* The Science Press, New York, 1929.

[Po,1951–6] Poincaré, H. J. *Oeuvres*, 11 vols. Gauthier-Villars, Paris, 1929.

[Po,1993] Poincaré, H. J. *New Methods of Celestial Mechanics.* Ed. and introduced by D. Goroff. American Institute of Physics, New York, 1993.

[P & S,1968] Pollard, H., and Saari, D. G. Singularities of the n-body problem. I. *Archive for Rational Mechanics and Analysis* **30** (1968), 263–269.

[P & S,1970] Pollard, H., and Saari, D. G. Singularities of the n-body problem. II. In *Inequalities—II*, pp. 255–259, ed. O. Shisha. Academic Press, New York, 1970.

[R,1985] Ratiu, T. Haretu's contribution to the n-body problem. *Libertas Mathematica* **5** (1985), 1–7.

[Re,1987] Regis, E. *Who Got Einstein's Office? Eccentricity and Genius at the Institute for Advanced Study.* Addison-Wesley, New York, 1987.

[Ru,1991] Ruelle, D. *Chance and Chaos.* Princeton University Press, Princeton, N.J., 1991.

[S,1971] Saari, D. G. Improbability of collisions in Newtonian gravitational systems. *Transactions of the American Mathematical Society* **162** (1971), 267–271; **168** (1972), 521; **181** (1973), 351–368.

[S,1973a] Saari, D. G. Singularities of Newtonian gravitational systems. In *Dynamical Systems, Salvador, Bahia, Brazil*, pp. 479–487, ed. M. M. Peixoto. Academic Press, New York, 1973.

[S,1973b] Saari, D. G. Singularities and collisions of Newtonian gravitational systems. *Archive for Rational Mechanics and Analysis* **49** (1973), 311–320.

[S,1973c] Saari, D. G. On global existence and uniqueness theorems for gravitational systems. *Celestial Mechanics* **8** (1973), 157–161.

[S,1975] Saari, D. G. Collisions are of first category. *Proceedings of the American Mathematical Society* **47**, 2 (1975), 442–445.

[S,1976] Saari, D. G. The n-body problem of celestial mechanics. *Celestial Mechanics* **14** (1976), 11–17.

[S,1977] Saari, D. G. A global existence and uniqueness theorem for the four-body problem of Newtonian mechanics. *Journal of Differential Equations* **26** (1977), 80–111.

[S,1978] Saari, D. G. Apportionment methods and the House of Representatives. *American Mathematical Monthly* **85** (1978), 792–802.

[S,1980] Saari, D. G. On the role and the properties of n-body central configurations. *Celestial Mechanics* **21** (1980), 9–20.

[S,1984] Saari, D. G. The manifold structure for collision and for hyperbolic-parabolic orbits in the *n*-body problem. *Journal of Differential Equations* **55** (1984), 300–329.

[S,1987] Saari, D. G. The source of some paradoxes from social choice and probability. *Journal of Economic Theory* **41** (1987), 1–22.

[S,1994] Saari, D. G. *Geometry of Voting*. Springer-Verlag, Berlin–Heidelberg–New York, 1994.

[S & X,1989] Saari, D. G., and Xia, Z. The existence of oscillatory and superhyperbolic motion in Newtonian systems. *Journal of Differential Equations* **82** (1989), 342–355.

[S & X,1995] Saari, D. G., and Xia, Z. Off to infinity in finite time. *Notices of the American Mathematical Society* **42** (1995), 538–546.

[Sh,1978] Sheldon, R. O. Noncollision singularities in the four-body problem. *Transactions of the American Mathematical Society* **249**, 2 (1979), 225–259.

[Si,1941] Siegel, C. L. Der Dreierstoss. *Annals of Mathematics* **42**, 1 (1941), 127–168.

[Si,1956] Siegel, C. L. *Vorlesungen über Himmelsmechanik*. Springer-Verlag, Berlin–Heidelberg–New York, 1956.

[S & M,1971] Siegel, C. L., and Moser, J. K. *Lectures on Celestial Mechanics*. Springer-Verlag, Berlin–Göttingen–Heidelberg, 1971.

[Sim,1977] Simó, C. Relative equilibrium solutions in the four-body problem. *Celestial Mechanics* **18** (1977), 165–184.

[Sim,1981] Simó, C. Analysis of triple collision in the isosceles problem. In *Classical Mechanics and Dynamical Systems*, pp. 203–224, ed. R. Devaney and Z. Nitecki. Lecture Notes in Pure and Applied Mathematics **70**. Dekker, New York, 1981.

[S & M, 1988] Simó, C., and Martinez, R. Qualitative study of the planar isosceles three-body problem. *Celestial Mechanics* **41** (1988), 179–251.

[Sit,1961] Sitnikov, K. The existence of oscillatory motion in the three-body problem. *Soviet Physics Doklady* **5** (1961), 647–650.

[Sm,1965] Smale, S. Diffeomorphisms with many periodic points. In *Differential and Combinatorial Topology: A Symposium in Honor of Marston Morse*, pp. 63–70, ed. S. S. Cairns. Princeton University Press, Princeton, N.J., 1965.

[Sm,1970a] Smale, S. Topology and mechanics II: The planar *n*-body problem. *Inventiones Mathematicae* **11** (1970), 45–64.

[Sm,1970b] Smale, S. Problems of the nature of relative equilibria in celestial mechanics. In *Manifolds-Amsterdam 1970*, pp. 194–198. Lecture Notes in Mathematics **197** (1970). Springer-Verlag, New York–Heidelberg–Berlin, 1975.

[Sm,1980] Smale, S. *The Mathematics of Time*. Springer-Verlag, New York, 1980.

[Sm,1991] An Hour's Conversation with Stephen Smale. *Nonlinear Science Today* **1**, 1 (1991), 1, 3, 12–18.

[Smi,1992] Smirnov, V. I. Biography of A. M. Lyapunov. *International Journal of Control* **55**, 3 (1992), 775–784.

[Sp,1970] Sperling, H. J. On the real singularities of the N-body problem. *Journal für die reine und angewandte Mathematik* **245** (1970), 15–40.

[St,1989] Stewart, I. *Does God Play Dice? The Mathematics of Chaos*. Basil Blackwell, Oxford, 1989.

[Su,1907] Sundman, K. Recherches sur le problème des trois corps. *Acta Societatis Scientiarum Fennicae* **34**, 6 (1907).

[Su,1909] Sundman, K. Nouvelles recherches sur le problème des trois corps. *Acta Societatis Scientiarum Fennicae* **35**, 9 (1909).

[Su,1912] Sundman, K. Mémoire sur le problème des trois corps. *Acta Mathematica* **36** (1912), 105–179.

[Sy,1933] Synge, J. L. John Charles Fields. *Journal of the London Mathematical Society* 8 (1933), 153–160.

[Sz,1967] Szebehely, V. Burrau's problem of the three bodies. *Proceedings of the National Academy of Sciences of the USA* **58** (1967), 60–65.

[Sz,1984] Szebehely, V. Review of concepts of stability. *Celestial Mechanics* **34** (1984), 49–64.

[T,1991–92] Tikhomirov, V. M., ed. *Selected Works of A. N. Kolmogorov*. Kluwer, Dordrecht–Boston–London: vol. 1, 1991; vol. 2, 1992.

[Tr,1976] Tropp, H. S. The origins and history of the Fields Medal. *Historia Mathematica* **3** (1976), 167–181.

[V & V,1927] van der Pol, B., and van der Mark, J. Frequency demultiplication. *Nature* **120** (1927), 363–364.

[V,1984] Verhulst, F. Perturbation theory from Lagrange to van der Pol. *Nieuw Archief voor Wiskunde* (4) **2** (1984), 428–438.

[W,1975] Waldvogel, J. The close triple approach. *Celestial Mechanics* **11** (1975), 429–432.

[W,1976] Waldvogel, J. The three-body problem near triple collision. *Celestial Mechanics* **14** (1976), 287–300.

[Wa,1993] Walters, P., ed. *Symbolic Dynamics and Its Applications*. Contemporary Mathematics, vol. 135. American Mathematical Society, Providence, R.I., 1993.

[We,1980] Westfall, R. *Never at Rest: A Biography of Isaac Newton*. Cambridge University Press, Cambridge, U.K., 1980.

[Wil,1913] Wilczynski, E. J. Ricerche geometriche intorno al problema dei tre corpi. *Annali di Matematica Pura ed Aplicata* **21** (1913), 1–31.

[Wi,1994] Wilson, R. Russian mathematics II. *The Mathematical Intelligencer* **16**, 3 (1994), 76.

[Win,1941] Wintner, A. *The Analytical Foundations of Celestial Mechanics*. Princeton University Press, Princeton, N.J., 1941.

[X,1991] Xia, Z. Central configurations with many small masses. *Journal of Differential Equations* **91** (1991), 168–179.

[X,1992] Xia, Z. The existence of noncollision singularities in the N-body problem. *Annals of Mathematics* **135** (1992), 411–468.

[X,1993] Xia, Z. Arnold diffusion in the elliptic restricted three-body problem. *Journal of Dynamics and Differential Equations* **5**, 2 (1993), 219–240.

[X,1994] Xia, Z. Arnold diffusion and oscillatory solutions in the planar three-body problem. *Journal of Differential Equations* **110** (1994), 289–321.

[Z,1987] Zdravkovska, S. Conversation with Vladimir Igorevich Arnold. *The Mathematical Intelligencer* **9**, 4 (1987), 28–32.

[Z & D,1994] Zdravkovska, S., and Duren, P. L., eds. *Golden Years of Moscow Mathematics*. History of Mathematics, vol. 6. American Mathematical Society, Providence, R.I., 1994.

[Ze,1993] Zehnder, E. Cantor-Medaille für Jürgen Moser. *Jahresbericht der deutschen Mathematiker Vereinigung* **95** (1993), 85–94.

[Zei,1908] Zeipel, H. von. Sur les singularités du problème des *n* corps. *Arkiv för Matematik, Astronomie och Fysik* **4**, 32 (1908), 1–4.

Index

Florin Diacu is Associate Professor of Mathematics at
the University of Victoria in Canada. Philip Holmes, a
Fellow of the American Academy of Arts and Sciences,
is Professor of Mechanics and Applied Mathematics at
Princeton University, where he directs the Program in
Applied and Computational Mathematics.